W9-CKR-783

THE LADY TASTING TEA

HOW STATISTICS REVOLUTIONIZED SCIENCE IN THE TWENTIETH CENTURY

DAVID SALSBURG

W. H. FREEMAN
AND COMPANY
NEW YORK

Cover design: Joan O'Conner
Text design: Victoria Tomaselli

Library of Congress Cataloging-in-Publication Data

Salsburg, David, 1931–
The lady tasting tea: how statistics revolutionized science in the twentieth century / by David Salsburg.
p. cm.
Includes bibliographical references and index.
ISBN 0-7167-4106-7 (hardcover)
1. Science—Statistical methods—History—20th century. I. Title.
Q175.S2345 2001
001.4'22'0904—dc21 00-049523

Printed in the United States of America
First printing 2001

W. H. Freeman and Company
41 Madison Avenue, New York, NY 10010
Houndmills, Basingstoke RG21 6XS, England

Dedicated to Fran, my dear wife of 42 years. As I accumulated stories about the men and women who made the statistical revolution during my career, she kept urging me to put them in a nonmathematical book. Fran, who has no mathematical training, helped me through the several revisions of this book, pointing out to me sections where my explanations were not clear. This book, especially those parts that are clearly understandable, is due to her persistence.

Thou shalt not answer questionnaires

Or quizzes upon World Affairs,

Nor with compliance

Take any test. Thou shalt not sit

With statisticians nor commit

A social science.

—W. H. AUDEN

To understand God's thoughts, we must study statistics, for these are the measure of his purpose.

—FLORENCE NIGHTINGALE

Contents

AUTHOR'S PREFACE

Science entered the nineteenth century with a firm philosophical vision that has been called the clockwork universe. It was believed that there were a small number of mathematical formulas (like Newton's laws of motion and Boyle's laws for gases) that could be used to describe reality and to predict future events. All that was needed for such prediction was a complete set of these formulas and a group of associated measurements that were taken with sufficient precision. It took over forty years for popular culture to catch up with this scientific vision.

Typical of this cultural lag is the exchange between Emperor Napoléon and Pierre Simon Laplace in the early years of the nineteenth century. Laplace had written a monumental and definitive book describing how to compute the future positions of planets and comets on the basis of a few observations from Earth. "I find no mention of God in your treatise, M. Laplace," Napoléon is reported to have said. "I had no need for that hypothesis," Laplace replied.

Many people were horrified with the concept of a godless clockwork universe running forever without divine intervention, with all future events determined by events of the past. In some sense, the romantic movement of the nineteenth century was a reaction to this cold, exact use of reasoning. However, a proof of this new science appeared in the 1840s, which dazzled the popular imagination. Newton's mathematical laws were used to predict the existence of another planet, and the planet Neptune was discovered in the place where these laws predicted it. Almost all resistance to the clockwork universe crumbled, and this philosophical stance became an essential part of popular culture.

However, if Laplace did not need God in his formulation, he did need something he called the "error function." The observations of planets and comets from this earthly platform did not fit the predicted positions exactly. Laplace and his fellow scientists attributed this to errors in the observations, sometimes due to perturbations in the earth's atmosphere, other times due to human error. Laplace threw all these errors into an extra piece (the error function) he tacked onto his mathematical descriptions. This error function sopped them up and left only the pure laws of motion to predict the true positions of celestial bodies. It was believed that, with more and more precise measurements, the need for an error function would diminish. With the error function to account for slight discrepancies between the observed and the predicted, early-nineteenth-century science was in the grip of philosophical determinism—the belief that everything that happens is determined in advance by the initial conditions of the universe and the mathematical formulas that describe its motions.

By the end of the nineteenth century, the errors had mounted instead of diminishing. As measurements became more and more precise, more and more error cropped up. The clockwork universe lay in shambles. Attempts to discover the laws of biology and sociology had failed. In the older sciences like physics and chemistry, the laws that Newton and Laplace had used were proving to be only rough approximations. Gradually, science began to work with a new paradigm, the statistical model of reality. By the end of the twentieth century, almost all of science had shifted to using statistical models.

Popular culture has failed to keep up with this scientific revolution. Some vague ideas and expressions (like "correlation," "odds," and "risk") have drifted into the popular vocabulary, and most people are aware of the uncertainties associated with some areas of science like medicine and economics, but few nonscientists have any understanding of the profound shift in philosophical view that has occurred. What are these statistical models?

How did they come about? What do they mean in real life? Are they true descriptions of reality? This book is an attempt to answer these questions. In the process, we will also look at the lives of some of the men and women who were involved in this revolution.

In dealing with these questions, it is necessary to distinguish among three mathematical ideas: randomness, probability, and statistics. To most people, randomness is just another word for unpredictability. An aphorism from the Talmud conveys this popular notion: "One should not go hunting for buried treasure, because buried treasure is found at random, and, by definition, one cannot go searching for something which is found at random." But, to the modern scientist, there are many different types of randomness. The concept of a probability distribution (which will be described in chapter 2 of this book) allows us to put constraints on this randomness and gives us a limited ability to predict future but random events. Thus, to the modern scientist, random events are not simply wild, unexpected, and unpredictable. They have a structure that can be described mathematically.

Probability is the current word for a very ancient concept. It appears in Aristotle, who stated that "It is the nature of probability that improbable things will happen." Initially, it involved a person's sense of what might be expected. In the seventeenth and eighteenth centuries, a group of mathematicians, among them the Bernoullis, a family of two generations, Fermat, de Moivre, and Pascal, all worked on a mathematical theory of probability, which started with games of chance. They developed some very sophisticated methods for counting equally probable events. de Moivre managed to insert the methods of calculus into these techniques, and the Bernoullis were able to discern some deep fundamental theorems, called "laws of large numbers." By the end of the nineteenth century, mathematical probability consisted primarily of sophisticated tricks but lacked a solid theoretical foundation.

In spite of the incomplete nature of probability theory, it proved useful in the developing idea of a statistical distribution. A statistical distribution occurs when we are considering a specific scientific problem. For instance, in 1971, a study from the Harvard School of Public Health was published in the British medical journal *Lancet,* which examined whether coffee drinking was related to cancer of the lower urinary tract. The study reported on a group of patients, some of whom had cancer of the lower urinary tract and some of whom had other diseases. The authors of the report also collected additional data on these patients, such as age, sex, and family history of cancer. Not everyone who drinks coffee gets urinary tract cancer, and not everyone who gets urinary tract cancer is a coffee drinker, so there are some events that contradict their hypothesis. However, 25 percent of the patients with this cancer habitually drank four or more cups of coffee a day. Only 10 percent of the patients without cancer were such heavy drinkers of coffee. There would seem to be some evidence in favor of the hypothesis.

This collection of data provided the authors with a statistical distribution. Using the tools of mathematical probability, they constructed a theoretical formula for that distribution, called the "probability distribution function," or just distribution function, which they used to examine the question. It is like Laplace's error function, but much more complicated. The construction of the theoretical distribution function makes use of probability theory, and it is used to describe what might be expected from future data taken at random from the same population of people.

This is not a book about probability and probability theory, which are abstract mathematical concepts. This is a book about the application of some of the theorems of probability to scientific problems, the world of statistical distributions, and distribution functions. Probability theory alone is insufficient to describe statistical methods, and it sometimes happens that statistical methods in science violate some of the theorems of probability. The reader will find probability drifting in and out of the chapters, being used where needed and ignored when not.

Because statistical models of reality are mathematical ones, they can be fully understood only in terms of mathematical formulas and symbols. This book is an attempt to do something a little less ambitious. I have tried to describe the statistical revolution in twentieth-century science in terms of some of the people (many of them still living) who were involved in that revolution. I have only touched on the work they created, trying to give the reader a taste of how their individual discoveries fit into the overall picture.

The reader of this book will not learn enough to engage in the statistical analysis of scientific data. That would require several years of graduate study. But I hope the reader will come away with some understanding of the profound shift in basic philosophy that is represented by the statistical view of science. So, where does the nonmathematician go to understand this revolution in science? I think that a good place to start is with a lady tasting tea. . . .

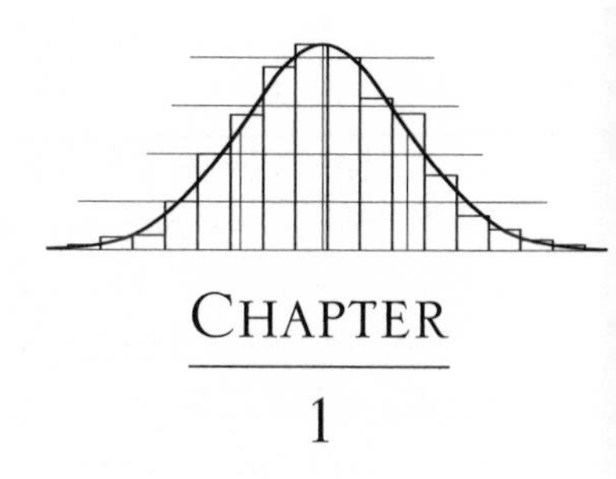

CHAPTER

1

THE LADY TASTING TEA

It was a summer afternoon in Cambridge, England, in the late 1920s. A group of university dons, their wives, and some guests were sitting around an outdoor table for afternoon tea. One of the women was insisting that tea tasted different depending upon whether the tea was poured into the milk or whether the milk was poured into the tea. The scientific minds among the men scoffed at this as sheer nonsense. What could be the difference? They could not conceive of any difference in the chemistry of the mixtures that could exist. A thin, short man, with thick glasses and a Vandyke beard beginning to turn gray, pounced on the problem.

"Let us test the proposition," he said excitedly. He began to outline an experiment in which the lady who insisted there was a difference would be presented with a sequence of cups of tea, in some of which the milk had been poured into the tea and in others of which the tea had been poured into the milk.

I can just hear some of my readers dismissing this effort as a minor bit of summer afternoon fluff. "What difference does it make whether the lady could tell one infusion from another?" they will ask. "There is nothing important or of great scientific merit in this problem," they will sneer. "These great minds should have been putting their immense brain power to something that would benefit mankind."

Unfortunately, whatever nonscientists may think about science and its importance, my experience has been that most scientists engage in their research because they are interested in the results and because they get intellectual excitement out of the work. Seldom do good scientists think about the eventual importance of their work. So it was that sunny summer afternoon in Cambridge. The lady might or might not have been correct about the tea infusion. The fun would be in finding a way to determine if she was right, and, under the direction of the man with the Vandyke beard, they began to discuss how they might make that determination.

Enthusiastically, many of them joined with him in setting up the experiment. Within a few minutes, they were pouring different patterns of infusion in a place where the lady could not see which cup was which. Then, with an air of finality, the man with the Vandyke beard presented her with her first cup. She sipped for a minute and declared that it was one where the milk had been poured into the tea. He noted her response without comment and presented her with the second cup....

THE COOPERATIVE NATURE OF SCIENCE

I heard this story in the late 1960s from a man who had been there that afternoon. He was Hugh Smith, but he published his scientific papers under the name H. Fairfield Smith. When I knew him, he was a professor of statistics at the University of Connecticut, in Storrs. I had received my Ph.D. in statistics from the University of Connecticut two years before. After teaching at the University of Pennsylvania, I had joined the clinical research department at Pfizer, Inc., a large pharmaceutical firm. Its research campus in

Groton, Connecticut, was about an hour's drive from Storrs. I was dealing with many difficult mathematical problems at Pfizer. I was the only statistician there at that time, and I needed to talk over these problems and my "solutions" to them.

What I had discovered working at Pfizer was that very little scientific research can be done alone. It usually requires a combination of minds. This is because it is so easy to make mistakes. When I would propose a mathematical formula as a means of solving a problem, the model would sometimes be inappropriate, or I might have introduced an assumption about the situation that was not true, or the "solution" I found might have been derived from the wrong branch of an equation, or I might even have made a mistake in arithmetic.

Whenever I would visit the university at Storrs to talk things over with Professor Smith, or whenever I would sit around and discuss problems with the chemists or pharmacologists at Pfizer, the problems I brought out would usually be welcomed. They would greet these discussions with enthusiasm and interest. What makes most scientists interested in their work is usually the excitement of working on a problem. They look forward to the interactions with others as they examine a problem and try to understand it.

THE DESIGN OF EXPERIMENTS

And so it was that summer afternoon in Cambridge. The man with the Vandyke beard was Ronald Aylmer Fisher, who was in his late thirties at the time. He would later be knighted Sir Ronald Fisher. In 1935, he wrote a book entitled *The Design of Experiments*, and he described the experiment of the lady tasting tea in the second chapter of that book. In his book, Fisher discusses the lady and her belief as a hypothetical problem. He considers the various ways in which an experiment might be designed to determine if she could tell the difference. The problem in designing the experiment is that, if she is given a single cup of tea, she has a 50 percent chance of guessing correctly which infusion was used, even if she cannot tell the difference. If she is given two cups of tea, she still might guess correctly. In fact, if she knew that the two cups of tea were each made with a different infusion, one guess could be completely right (or completely wrong).

Similarly, even if she could tell the difference, there is some chance that she might have made a mistake, that one of the cups was not mixed as well or that the infusion was made when the tea was not hot enough. She might be presented with a series of ten cups and correctly identify only nine of them, even if she could tell the difference.

In his book, Fisher discusses the various possible outcomes of such an experiment. He describes how to decide how many cups should be presented and in what order and how much to tell the lady about the order of presentations. He works out the probabilities of different outcomes, depending upon whether the lady is or is not correct. Nowhere in this discussion does he indicate that such an experiment was ever run. Nor does he describe the outcome of an actual experiment.

The book on experimental design by Fisher was an important element in a revolution that swept through all fields of science in the first half of the twentieth century. Long before Fisher came on the scene, scientific experiments had been performed for hundreds of years. In the later part of the sixteenth century, the English physician William Harvey experimented with animals, blocking the flow of blood in different veins and arteries, trying to trace the circulation of blood as it flowed from the heart to the lungs, back to the heart, out to the body, and back to the heart again.

Fisher did not discover experimentation as a means of increasing knowledge. Until Fisher, experiments were idiosyncratic to each scientist. Good scientists would be able to construct experiments that produced new knowledge. Lesser scientists would often engage in "experimentation" that accumulated much data but was useless for increasing knowledge. An example of this can be seen in the many inconclusive attempts that were made during the late nineteenth century to measure the speed of light. It was not until the American physicist Albert Michelson constructed a highly sophisticated series of experiments with light and mirrors that the first good estimates were made.

In the nineteenth century, scientists seldom published the results of their experiments. Instead, they described their conclusions and published data that "demonstrated" the truth of those conclusions. Gregor Mendel did not show the results of all his experiments in breeding peas. He described the sequence of experiments and then wrote: "The first ten members of both series of experiments may serve as an illustration. . . ." (In the 1940s, Ronald Fisher examined Mendel's "illustrations" of data and discovered that the data were too good to be true. They did not display the degree of randomness that should have occurred.)

Although science has been developed from careful thought, observations, and experiments, it was never quite clear how one should go about experimenting, nor were the complete results of experiments usually presented to the reader.

This was particularly true for agricultural research in the late nineteenth and early twentieth centuries. The Rothamsted Agricultural Experimental Station, where Fisher worked during the early years of the twentieth century, had been experimenting with different fertilizer components (called "artificial manures") for almost ninety years before he arrived. In a typical experiment, the workers would spread a mixture of phosphate and nitrogen salts over an entire field, plant grain, and measure the size of the harvest, along with the amount of rainfall during that summer. There were elaborate formulas used to "adjust" the output of one year or one field, in order to compare it to the output of another field or of the same field in another year. These were called "fertility indexes," and each agricultural experimental station had its own fertility index, which it believed was more accurate than any other.

The result of these ninety years of experimentation was a mess of confusion and vast troves of unpublished and useless data. It seemed as if some strains of wheat responded better than other strains to one fertilizer, but only in years when rainfall was excessive. Other experiments seemed to show that sulfate of potash one year, followed by sulfate of soda for the next year, produced an increase in some varieties of potatoes but not others. The most that could be said of these artificial manures was that some of them worked sometimes, perhaps, or maybe.

Fisher, a consummate mathematician, looked at the fertility index that the agricultural scientists at Rothamsted used to correct the results of experiments to account for differences due to the weather from year to year. He examined the competing indexes used by other agricultural experimental stations. When reduced to their elemental algebra, they were all versions of the same formula. In other words, two indexes, whose partisans were hotly contending, were really making exactly the same correction. In 1921, he published a paper in the leading agricultural journal, the *Annals of Applied Biology*, in which he showed that it did not make any difference what index was used. The article also showed that all these corrections were inadequate to adjust for differences in the fertility of different fields. This remarkable paper ended over twenty years of scientific dispute.

Fisher then examined the data on rainfall and crop production over the previous ninety years and showed that the effects of different weather from year to year were far greater than any effect of different fertilizers. To use a word Fisher developed later in his theory of experimental design, the year-to-year differences in weather and the year-to-year differences in artificial manures were "confounded." This means that there was no way to pull them apart using data from these experiments. Ninety years of experimentation and over twenty years of scientific dispute had been an almost useless waste of effort!

This set Fisher thinking about experiments and experimental design. He concluded that the scientist needs to start with a mathematical model of the outcome of the potential experiment. A mathematical model is a set of equations, in which some of the symbols stand for numbers that will be collected as data from the experiments and other symbols stand for the overall outcomes of the experiment. The scientist starts with the data from the experiment and computes outcomes appropriate to the scientific question being considered.

Consider a simple example from the experience of a teacher with a particular student. The teacher is interested in finding some measure of how much the child has learned. To this end, the teacher "experiments" by giving the child a group of tests. Each test is marked on a scale from 0 to 100. Any one test provides a poor estimate of how much the child knows. It may be that the child did not study the few things that were on that test but knows a great deal about things that were not on the test. The child may have had a headache the day she took a particular test. The child may have had an argument with parents the morning of a particular test. For many reasons, one test does not provide a good estimate of knowledge. So, the teacher gives a set of tests. The average score from all those tests is taken as a better estimate of how much the child knows. How much the child knows is the outcome. The scores on individual tests are the data.

How should the teacher structure those tests? Should they be a sequence of tests that cover only the material taught over the past couple of days? Should they each involve something from all the material taught until now? Should the tests be given weekly, or daily, or at the end of each unit being taught? All of these are questions involved in the design of the experiment.

When the agricultural scientist wants to know the effect of a particular artificial fertilizer on the growth of wheat, an experiment has to be constructed that will provide data to estimate that effect. Fisher showed that the first step in the design of that experiment is to set up a group of mathematical equations describing the relationship between the data that will be collected and the outcomes that are being estimated. Then, any useful experiment has to be one that allows for estimation of those outcomes. The experiment has to be specific and enable the scientist to determine the difference in outcome that is due to weather versus the difference that is due to the use of different fertilizers. In particular, it is necessary to include all the treatments being compared in the same experiment, something that came to be called "controls."

In his book, *The Design of Experiments*, Fisher provided a few examples of good experimental designs, and derived general rules for good designs. However, the mathematics involved in Fisher's methods were very complicated, and most scientists were unable to generate their own designs unless they followed the pattern of one of the designs Fisher derived in his book.

Agricultural scientists recognized the great value of Fisher's work on experimental design, and Fisherian methods were soon dominating schools of agriculture in most of the English-speaking world. Taking off from Fisher's initial work, an entire body of scientific literature has developed to describe different experimental designs. These designs have been applied to fields other than agriculture, including medicine, chemistry, and industrial quality control. In many cases, the mathematics involved are deep and complicated. But, for the moment, let us stop with the idea that the scientist cannot just go off and "experiment." It takes some long and careful thought—and often a strong dose of difficult mathematics.

And the lady tasting tea, what happened to her? Fisher does not describe the outcome of the experiment that sunny summer afternoon in Cambridge. But Professor Smith told me that the lady identified every single one of the cups correctly.

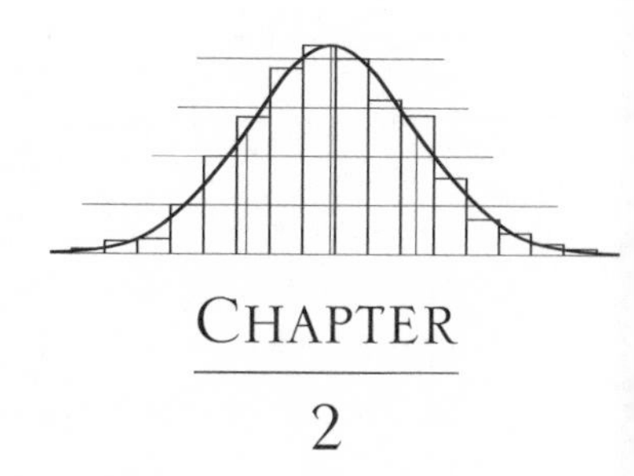

CHAPTER

2

THE SKEW DISTRIBUTIONS

As with many revolutions in human thought, it is difficult to find the exact moment when the idea of a statistical model became part of science. One can find possible specific examples of it in the work of the German and French mathematicians of the early nineteenth century, and there is even a hint of it in the papers of Johannes Kepler, the great seventeenth-century astronomer. As indicated in the preface to this book, Laplace invented what he called the error function to account for statistical problems in astronomy. I would prefer to date the statistical revolution to the work of Karl Pearson in the 1890s. Charles Darwin recognized biological variation as a fundamental aspect of life and made it the basis of his theory of the survival of the fittest. But it was his fellow Englishman Karl Pearson who first recognized the underlying nature of statistical models and how they offered something different from the deterministic view of nineteenth-century science.

When I began the study of mathematical statistics in the 1960s, Pearson was seldom mentioned in my classes. As I met and talked with the major figures in the field, I heard no references to Pearson or his work. He was either ignored or treated as a minor figure whose activities had long since been outmoded. Churchill Eisenhart, from the U.S. National Bureau of Standards, for instance, was studying at University College, London, during the final years of Karl Pearson's life. He remembered Pearson as a dispirited old man. The pace of statistical research had swept him by, dashing him and most of his work into the dustbin of the past. The bright young students at University College were flocking to study at the feet of the newer great men, one of them Karl Pearson's own son, but no one was coming to see old Karl in his lonely office far from the bustle of new, exciting research.

It wasn't always like this. In the 1870s, young Carl [sic] Pearson had left England to pursue his graduate studies in political science in Germany. There he became enamored of the work of Karl Marx. In tribute to Marx, he changed the spelling of his own first name to Karl. He returned to London with a doctorate in political science, having written two respectable books in the field. In the very heart of stuffy Victorian England, he had the audacity to organize a Young Men's and Women's Discussion Club. At the club, young men and women gathered together (unchaperoned), in an equality of the sexes modeled after the salons of upper-class German and French society. There they discussed the great political and philosophical problems of the world. The fact that Pearson met his wife in this environment suggests that there may have been more than one motive for founding the club. This little social venture provides some insight into Karl Pearson's original mind and his utter disregard for established tradition.

Although his doctorate was in political science, Pearson's main interests were in the philosophy of science and the nature of mathematical modeling. In the 1880s, he published *The Grammar of Science*, which went through a number of editions. For much of the period prior to World War I, this was considered one of the great books on the nature of science and mathematics. It is filled

Karl Pearson, 1857–1936, with a bust of Raphael Weldon in the background

with brilliant, original insights, which make it an important work in the philosophy of science. It was also written in a smooth, simple style that makes it accessible to anyone. You don't have to know mathematics to read and understand the *Grammar of Science*. Although, at this writing, the book is over a hundred years old, the insights and the ideas found in it are pertinent to much mathematical research of the twenty-first century and provide an understanding of the nature of science that holds true even today.

The Galton Biometrical Laboratory

At this point in his life, Pearson fell under the influence of the English scientist Sir Francis Galton. Most people who have heard of Galton know him as the "discoverer" of fingerprints. The realization that fingerprints are unique to each individual and the methods usually used to classify and identify them are Galton's. The uniqueness of fingerprints lies in the occurrence of irregular marks and cuts in the finger patterns, which are called "Galton Marks." Galton did far more. Independently wealthy, he was a dilettante scientist who sought to bring mathematical rigor into the science of biology through the study of patterns of numbers. One of his first investigations involved the inheritance of genius. He collected information about pairs of fathers and sons who were reputed to be highly intelligent. He found the problem very difficult, however, because there was no good measure of intelligence at the time. He decided to look at the inheritance of traits that were more easily measured, like height.

Galton set up a biometrical laboratory (*bio* for biology, *metric* for measurement) in London and advertised for families to come and be measured. At the biometrical laboratory, he collected heights, weights, measurements of specific bones, and other characteristics of family members. He and his assistants tabulated these data and examined and reexamined them. He was looking for some way to predict measures from parents to children. It was obvious, for instance, that tall parents tended to have tall children, but was there some mathematical formula that would predict how tall the children would be, using only the heights of the parents?

CORRELATION AND REGRESSION

In this way, Galton discovered a phenomenon he called "regression to the mean." It turned out that sons of very tall fathers tended to be shorter than their fathers and sons of very short fathers tended to be taller than their fathers. It was as if some mysterious force were causing human heights to move away from the extremes and toward the mean or average of all humans. The phenomenon of regression to the mean holds for more than human heights. Almost all scientific observations are bedeviled by regression to the mean. We shall see in chapters 5 and 7 how R. A. Fisher was able to turn Galton's regression to the mean into statistical models that now dominate economics, medical research, and much of engineering.

Galton thought about his remarkable finding and then realized that it had to be true, that it could have been predicted before making all his observations. Suppose, he said, that regression to the mean did not occur. Then, on the average, the sons of tall fathers would be as tall as their fathers. In this case, some of the sons would have to be taller than their fathers (in order to average out the ones who are shorter). The sons of this generation of taller men would then average their heights, so some sons would be even taller. It would go on, generation after generation. Similarly, there would be some sons shorter than their fathers, and some grandsons even shorter, and so on. After not too many generations, the human race would consist of ever taller people at one end and ever shorter ones at the other.

This does not happen. The heights of humans tend to remain stable, on the average. This can only happen if the sons of very tall fathers average shorter heights and the sons of very short fathers average greater heights. Regression to the mean is a phenomenon that maintains stability and keeps a given species pretty much the "same" from generation to generation.

Galton discovered a mathematical measure of this relationship. He called it the "coefficient of correlation." Galton gave a specific formula for computing this number from the type of data he collected at the biometrical laboratory. It is a highly specific formula for measuring one aspect of regression to the mean, but tells us nothing whatsoever about the cause of that phenomenon. Galton first used the word *correlation* in this sense. It has since moved into popular language. Correlation is often used to mean something much more vague than Galton's specific coefficient of correlation. It has a scientific ring to its sound, and nonscientists often bandy the word around as if it described how two things are related. But unless you are referring to Galton's mathematical measure, you are not being very precise or scientific if you use the word correlation, which Galton used for this specific purpose.

DISTRIBUTIONS AND PARAMETERS

With the formula for correlation, Galton was getting very close to this new revolutionary idea that was to modify almost all science in the twentieth century. But it was his disciple, Karl Pearson, who first formulated the idea in its most complete form.

To understand this revolutionary idea, you have to cast aside all preconceived notions about science. Science, we are often taught, is measurement. We make careful measurements and use them to find mathematical formulas that describe nature. In high school physics, we are taught that the distance a falling body will travel versus time is given by a formula involving a symbol **g**, where **g** is the constant of acceleration. We are taught that experiments can be used to determine the value of **g**. Yet, when the high school student runs a sequence of experiments to determine the value of **g**, rolling small weights along an inclined plane and measuring how long it takes them to get to different places on the ramp, what happens? It seldom comes out right. The more times the student runs the experiment, the more confusion occurs, as different values of **g** emerge from different experiments. The teacher looks down from his superior knowledge and assures the students that they are not getting the right answer because they are sloppy or not being careful or have copied incorrect numbers.

What he does not tell them is that all experiments are sloppy and that very seldom does even the most careful scientist get the number right. Little unforeseen and unobservable glitches occur in every experiment. The air in the room might be too warm and the sliding weight might stick for a microsecond before it begins to slide. A slight breeze from a passing butterfly might have an effect. What one really gets out of an experiment is a scatter of numbers, not one of which is right but all of which can be used to get a close estimate of the correct value.

Armed with Pearson's revolutionary idea, we do not look upon experimental results as carefully measured numbers in their own right. Instead, they are examples of a scatter of numbers, a *distribution* of numbers, to use the more accepted term. This distribution of numbers can be written as a mathematical formula that tells us the probability that an observed number will be a given value. What value that number actually takes in a specific experiment is unpredictable. We can only talk about probabilities of values and not about certainties of values. The results of individual experiments are random, in the sense that they are unpredictable. The statistical models of distributions, however, enable us to describe the mathematical nature of that randomness.

It took some time for science to realize the inherent randomness of observations. In the eighteenth and nineteenth centuries, astronomers and physicists created mathematical formulas that described their observations to a degree of accuracy that was acceptable. Deviations between observed and predicted values were expected because of the basic imprecision of the measuring instruments, and were ignored. Planets and other astronomical bodies were assumed to follow precise paths determined by the fundamental equations of motion. Uncertainty was due to poor instrumentation. It was not inherent in nature.

With the development of ever more precise measuring instruments in physics, and with attempts to extend this science of measurement to biology and sociology, the inherent randomness of nature became more and more clear. How could this be handled? One way was to keep the precise mathematical formulas and treat the deviations between the observed values and the predicted values as a small, unimportant error. In fact, as early as 1820, mathematical papers by Laplace describe the first probability distribution, the error distribution, that is a mathematical formulation of the probabilities associated with these small, unimportant errors.

This error distribution has entered popular parlance as the "bell-shaped curve," or the normal distribution.[1]

It took Pearson to go one step beyond the normal, or error, distribution. Looking at the data accumulated in biology, Pearson conceived of the measurements themselves, rather than errors in the measurement, as having a probability distribution. Whatever we measure is really part of a random scatter, whose probabilities are described by a mathematical function, the distribution function. Pearson discovered a family of distribution functions that he called the "skew distributions" and that, he claimed, would describe any type of scatter a scientist might see in data. Each of the distributions in this family is identified by four numbers.

The numbers that identify the distribution function are not the same type of "number" as the measurements. These numbers can never be observed but can be inferred from the way in which the measurements scatter. These numbers were later to be called parameters—from the Greek for "almost measurements." The four parameters that completely describe a member of the Pearson System are called

1. the mean—the central value about which the measurements scatter,
2. the standard deviation—how far most of the measurements scatter about the mean,
3. symmetry—the degree to which the measurements pile up on only one side of the mean,
4. kurtosis—how far rare measurements scatter from the mean.

[1]It is sometimes called the Gaussian distribution, in honor of the man once believed to have first formulated it, except that it was not Carl Friedrich Gauss but an earlier mathematician named Abraham de Moivre who first wrote down the formula for the distribution. There is good reason to believe Daniel Bernoulli came across the formula before this. All of this is an example of what Stephen Stigler, a contemporary historian of science, calls the law of misonomy, that nothing in mathematics is ever named after the person who discovered it.

There is a subtle shift in thinking with Pearson's system of skew distributions. Before Pearson, the "things" that science dealt with were real and palpable. Kepler attempted to discover the mathematical laws that described how the planets moved in space. William Harvey's experiments tried to determine how blood moved through the veins and arteries of a specific animal. Chemistry dealt with elements and compounds made up of elements. However, the "planets" that Kepler tried to tame were really a set of numbers identifying the positions in the sky where shimmering lights were seen by observers on earth. The exact course of blood through the veins of a single horse was different from what might have been seen with a different horse, or with a specific human being. No one was able to produce a pure sample of iron, although it was known to be an element.

Pearson proposed that these observable phenomena were only random reflections. What was real was the probability distribution. The real " things" of science were not things that we could observe and hold but mathematical functions that described the randomness of what we could observe. The four parameters of a distribution are what we really want to determine in a scientific investigation. In some sense, we can never really determine those four parameters. We can only estimate them from the data.

Pearson failed to recognize this last distinction. He believed that if we collected enough data the estimates of the parameters would provide us with true values of the parameters. It took his younger rival, Ronald Fisher, to show that many of Pearson's methods of estimation were less than optimal. In the late 1930s, as Karl Pearson was approaching the end of his long life, a brilliant young Polish mathematician, Jerzy Neyman, showed that Pearson's system of skew distributions did not cover the universe of possible distributions and that many important problems could not be solved using the Pearson system.

But let us leave the old, abandoned Karl Pearson of 1934 and return to the vigorous man in his late thirties, who was filled with enthusiasm over his discovery of skew distributions. In 1897, he took over Galton's biometrical laboratory in London and marshaled legions of young women (called "calculators") to compute the parameters of distributions associated with the data Galton had been accumulating on human measurements. At the turn of the new century, Galton, Pearson, and Raphael Weldon combined their efforts to found a new scientific journal that would apply Pearson's ideas to biological data. Galton used his wealth to create a trust fund that would support this new journal. In the first issue, the editors set forth an ambitious plan.

THE PLAN OF *BIOMETRIKA*

Galton, Pearson, and Weldon were part of an exciting cadre of British scientists who were exploiting the insights of one of their most prominent members, Charles Darwin. Darwin's theories of evolution postulated that life forms change in response to environmental stress. He proposed that changing environments gave a slight advantage to those random changes that fit better into the new environment. Gradually, as the environment changed and life forms continued to have random mutations, a new species would emerge that was better fit to live and procreate in the new environment. This idea was given the shorthand designation "survival of the fittest." It had an unfortunate effect on society when arrogant political scientists adapted it to social life, declaring that those who emerged triumphant from the economic battle over riches were more fit than those who plunged into poverty. Survival of the fittest became a justification for rampant capitalism in which the rich were given the moral authority to ignore the poor.

In the biological sciences, Darwin's ideas seemed to have great validity. Darwin could point to the resemblances among related species as suggesting a previous species out of which these modern ones had emerged. Darwin showed how small birds of slightly different species and living on isolated islands had many anatomical commonalities. He pointed to the similarities among embryos of different species, including the human embryo, which starts with a tail.

The one thing Darwin was unable to show was an example of a new species actually emerging within the time frame of human history. Darwin postulated that new species emerge because of the survival of the fittest, but there was no proof of this. All he had to display were modern species that appeared to "fit" well within their environment. Darwin's proposals seemed to account for what was known, and they had an attractive logical structure to them. But, to translate an old Yiddish expression, "For instance is no proof."

Pearson, Galton, and Weldon set out in their new journal to rectify this. In Pearson's view of reality as probability distributions, Darwin's finches (an important example he used in his book) were not the objects of scientific investigation. The random distribution of all finches of a species was the object. If one could measure the beak lengths of all the finches in a given species, the distribution function of those beak lengths would have its own four parameters, and those four parameters would *be* the beak length of the species.

Suppose, Pearson said, that there was an environmental force changing a given species by providing superior survivorship to certain specific random mutations. We might not be able to live long enough to see a new species emerge, but we might be able to see a change in the four parameters of the distribution. In their first issue, the three editors declared that their new journal would collect data from all over the world and determine the parameters of their distributions, with the eventual hope of showing examples of shifts in parameters associated with environmental change.

They named their new journal *Biometrika*. It was funded lavishly by the Biometrika Trust that Galton set up, and was so well funded that it was the first journal to publish full color photographs and foldout glassine sheets with intricate drawings. It was printed on high-quality rag paper, and the most complicated mathematical formulas were displayed, even if they meant extremely complicated and expensive typesetting.

For the next twenty-five years, *Biometrika* printed data from correspondents who plunged into the jungles of Africa to measure tibia and fibula of the natives; sent in beak lengths of exotic tropical birds caught in the rain forests of Central America; or raided ancient cemeteries to uncover human skulls, into which they poured buckshot to measure cranial capacity. In 1910, the journal published several sheets of full color photographs of flaccid penises of pygmy men, laid on a flat surface against measuring sticks.

In 1921, a young female correspondent, Julia Bell, described the troubles she underwent when she tried to get anthropomorphic measurements of recruits for the Albanian army. She left Vienna for a remote outpost in Albania, assured that she would find German-speaking officers to help her. When she arrived, there was only a sergeant, who spoke three words of German. Undaunted, she took out her bronze measuring rods and got the young men to understand what she wanted by tickling them until they lifted their arms or legs as she desired.

For each of these data sets, Pearson and his calculators computed the four parameters of the distributions. The articles would display a graphical version of the best-fitting distribution and some comments about how this distribution differed from the distribution of other related data. In retrospect, it is difficult to see how all this activity helped prove Darwin's theories. Reading through these issues of *Biometrika*, I get the impression that it soon became an effort that was done for its own sake and had no real purpose other than estimating parameters for a given set of data.

Scattered throughout the journal are other articles. Some of them involve theoretical mathematics dealing with problems that arise from the development of probability distributions. In 1908, for instance, an unknown author, publishing under the pseudonym of "Student," produced a result that plays a role in almost all modern scientific work, "Student"'s t-test. We will meet this anonymous author in later chapters and discuss his unfortunate role in mediating between Karl Pearson and Ronald Fisher.

Galton died in 1911, and Weldon died in a skiing accident in the Alps earlier. This left Pearson as the sole editor of *Biometrika* and the sole dispenser of the trust's money. In the next twenty years, it was Pearson's personal journal, which published what Pearson thought was important and did not publish what Pearson thought was unimportant. It was filled with editorials written by Pearson, in which he let his fertile imagination range over all sorts of issues. Renovation of an ancient Irish church uncovered bones in the walls, and Pearson used involved mathematical reasoning and measurements made on those bones to determine whether they were, in fact, the bones of a particular medieval saint. A skull was found that was purported to be the skull of Oliver Cromwell. Pearson investigated this in a fascinating article that described the known fate of Cromwell's body, and then compared measurements made on pictures painted of Cromwell to measurements made on the skull.[2] In other articles, Pearson examined the lengths of reigns of kings and the decline of the patrician class in ancient Rome, and made other forays into sociology, political science, and botany, all of them with a complicated mathematical gloss.

Just before his death, Karl Pearson published a short article entitled "On Jewish–Gentile Relationships," in which he analyzed anthropomorphic data on Jews and Gentiles from various parts of the world. He concluded that the racial theories of the National Socialists, the official name of the Nazis, were sheer nonsense, that there was no such thing as a Jewish race or, for that matter, an Aryan race. This final paper was well within the clear, logical, carefully reasoned tradition of his previous work.

[2]After the restoration of the monarchy, following Cromwell's dictatorship, a truce between the two factions in the civil war in England meant that the new rulers could not prosecute any of the living followers of Cromwell. However, there was nothing in the truce about the dead. So the bodies of Cromwell and two of the judges who had ordered the execution of Charles I were dug up and tried for the crime of regicide. They were convicted, and their heads were chopped off and placed on pikes above Westminster Abbey. The three heads were left there for years and eventually disappeared. A head, supposedly that of Cromwell, showed up in a "museum" in London. It was that head which Pearson examined. He concluded that it was, indeed, the head of Oliver Cromwell.

Pearson used mathematics to investigate many areas of human thought that few would consider the normal business of science. To read through his editorials in *Biometrika* is to meet a man with a universal range of interests and a fascinating capacity to cut to the heart of any problem and find a mathematical model with which to attack it. To read through his editorials is also to meet a strong-willed, highly opinionated man, who viewed subordinates and students as extensions of his own will. I think I would have enjoyed spending a day with Karl Pearson—provided I did not have to disagree with him.

Did they prove Darwin's theory of evolution through survival of the fittest? Perhaps they did. By comparing the distributions of cranial capacity from skulls in ancient cemeteries to those of modern men and women, they managed to show that the human species has been remarkably stable across many thousands of years. By showing that anthropomorphic measurements on aborigines had the same distribution as measurements taken on Europeans, they disproved the claims of some Australians that the aborigines were not human. Out of this work, Pearson developed a basic statistical tool known as the "goodness of fit test," which is an indispensable tool for modern science. It enables the scientist to determine whether a given set of observations is appropriate to a particular mathematical distribution function. In chapter 10, we shall see how Pearson's own son used this goodness of fit test to undermine much of what his father had accomplished.

As the twentieth century advanced, more and more of the articles in *Biometrika* dealt with theoretical problems in mathematical statistics and fewer dealt with distributions of specific data. When Karl Pearson's son, Egon Pearson, took over as editor, the shift to theoretical mathematics was complete, and today *Biometrika* is a preeminent journal in that field.

But did they prove survival of the fittest? The closest they came to it occurred early in the twentieth century. Raphael Weldon conceived of a grand experiment. The development of china factories in southern England in the eighteenth century had caused some of the rivers to become silted with clay, so the harbors of Plymouth and Dartmouth had changed, with the interior regions more silted than those closer to the sea. Weldon took several hundred crabs from these harbors and put them into individual glass jars. In half the jars he used the silted water from the inner harbors. In the other half of the jars he used clearer water from the outer harbors. He then measured the carapaces of the crabs that survived after a period of time and determined the parameters of the two distributions of crabs: those that survived in clear water and those that survived in silted water.

Just as Darwin had predicted, the crabs that survived in the silted jars showed a change in distribution parameters! Did this prove the theories of evolution? Unfortunately, Weldon died before he could write up the results of his experiment. Pearson described the experiment and its results in a preliminary analysis of the data. A final analysis was never run. The British government, which had supplied the funds for the experiment, demanded a final report. The final report never came. Weldon was dead, and the experiment was ended.

Eventually, Darwin's theories were shown to be true for short-lived species like bacteria and fruit flies. Using these species, the scientist could experiment with thousands of generations in a short interval of time. Modern investigations of DNA, the building blocks of heredity, have provided even stronger evidence of the relationships among species. If we assume that the rate of mutation has been constant over the past ten million or more years, studies of DNA can be used to estimate the time frame of species emergence for primates and other mammals. At a minimum, it runs into the hundreds of thousands of years. Most scientists now accept Darwin's mechanism of evolution as correct. No other theoretical mechanism has been proposed that matches all known data so well. Science is satisfied, and the idea that one needs to determine the shift in distribution parameters to show evolution on a short time scale has been dropped.

What remains of the Pearsonian revolution is the idea that the "things" of science are not the observables but the mathematical distribution functions that describe the probabilities associated with observations. Today, medical investigations use subtle mathematical models of distributions to determine the possible effects of treatments on long-term survival. Sociologists and economists use mathematical distributions to describe the behavior of human society. In the form of quantum mechanics, physicists use mathematical distributions to describe subatomic particles. No aspect of science has escaped the revolution. Some scientists claim that the use of probability distributions is a temporary stopgap and that, eventually, we will be able to find a way to return to the determinism of nineteenth-century science. Einstein's famous dictum that he did not believe that the Almighty plays dice with the universe is an example of that view. Others believe that nature is fundamentally random and that the only reality lies in distribution functions. Regardless of one's underlying philosophy, the fact remains that Pearson's ideas about distribution functions and parameters came to dominate twentieth-century science and stand triumphant on the threshold of the twenty-first century.

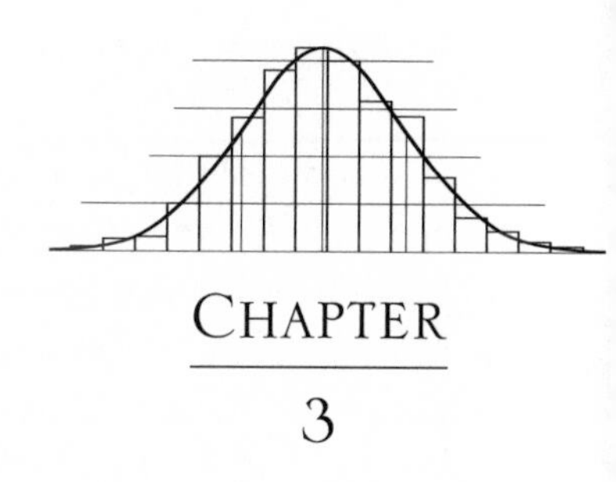

CHAPTER 3

THAT DEAR MR. GOSSET

That ancient and honorable firm, the Guinness Brewing Company of Dublin, Ireland, began the twentieth century with an investment in science. Young Lord Guinness had recently inherited the enterprise, and he decided to introduce modern scientific techniques into the business by hiring the leading graduates in chemistry from Oxford and Cambridge Universities. In 1899, he recruited William Sealy Gosset, who had just graduated from Oxford at age twenty-three with a combined degree in chemistry and mathematics. Gosset's mathematical background was a traditional one of the time, including calculus, differential equations, astronomy, and other aspects of the clockwork universe view of science. The innovations of Karl Pearson and the first glimmerings of what was to become quantum mechanics had not yet made their way into the university curriculum. Gosset had been hired for his expertise in chemistry. What use would a brewery have for a mathematician?

Gosset turned out to be a good investment for Guinness. He showed himself to be a very able administrator, and he eventually rose in the company to be in charge of its entire Greater London operations. It was, in fact, as a mathematician that he made his first major contribution to the art of brewing beer. A few years before, the Danish telephone company was one of the first industrial companies to hire a mathematician, but they had a clearly mathematical problem: how big to make the switchboard of a telephone exchange. Where in the making of beer and ale would there be a mathematical problem to be solved?

Gosset's first published paper, in 1904, deals with such a problem. When the mash was prepared for fermentation, a carefully measured amount of yeast was used. Yeast are living organisms, and cultures of yeast were kept alive and multiplying in jars of fluid before being put into the mash. The workers had to measure how much yeast was in a given jar in order to determine how much fluid to use. They drew a sample of the fluid and examined it under a microscope, counting the number of yeast cells they saw. How accurate was that measure? It was important to know, because the amount of yeast used in the mash had to be carefully controlled. Too little led to incomplete fermentation. Too much led to bitter beer.

Note how this matches Pearson's approach to science. The measurement was the count of yeast cells in the sample, but the real "thing" that was sought was the concentration of yeast cells in the entire jar. Since the yeast was alive, and cells were constantly multiplying and dividing, this "thing" did not really exist. What did exist in some sense was the probability distribution of yeast cells per unit volume. Gosset examined the data and determined that the counts of yeast cells could be modeled with a probability distribution known as the "Poisson distribution."[1] This was not one of Pearson's family of skew distributions. In fact, it is a peculiar distribution that has only one parameter (instead of four).

[1]Following Stigler's law of misonomy, the Poisson distribution is named after the eighteenth–nineteenth-century mathematician Siméon Denis Poisson, but the distribution named after him was described earlier by one of the Bernoullis.

Having determined that the number of live yeast cells in a sample follows a Poisson distribution, Gosset was able to devise rules and methods of measurement that led to much more accurate assessments of the concentration of yeast cells. Using Gosset's methods, Guinness was able to produce a much more consistent product.

THE BIRTH OF "STUDENT"

Gosset wanted to publish this result in an appropriate journal. The Poisson distribution (or the formula for it) had been known for over 100 years, and attempts had been made in the past to find examples of it in real life. One such attempt involved counting the number of soldiers in the Prussian Army who died from horse kicks. In his yeast cell counts, Gosset had a clear example, along with an important application of the new idea of statistical distributions. However, it was against company policy to allow publications by its employees. A few years before, a master brewer from Guinness had written an article in which he revealed the secret components of one of their brewing processes. To avoid the further loss of such valuable company property, Guinness had forbidden its employees from publishing.

Gosset had become friendly with Karl Pearson, one of the editors of *Biometrika* at the time, and Pearson was impressed with Gosset's great mathematical ability. In 1906, Gosset convinced his employers that the new mathematical ideas were useful for a beer company and took a one-year leave of absence to study under Pearson at the Galton Biometrical Laboratory. Two years before this, when Gosset described his results dealing with yeast, Pearson was eager to print it in his journal. They decided to publish the article using a pseudonym. This first discovery of Gosset's was published by an author identified only as "Student."

Over the next thirty years, "Student" wrote a series of extremely important papers, almost all of them printed in *Biometrika*. At some point, the Guinness family discovered that their "dear Mr. Gosset" had been secretly writing and publishing scientific papers, contrary to company policy. Most of "Student"'s mathematical activity took place at home, after normal working hours, and his rise in the company to ever more responsible positions shows that the Guinness company was not ill-served by Gosset's extracurricular work. There is an apocryphal story that the first time the Guinness family heard of this work occurred when Gosset died suddenly of a heart attack in 1937 and his mathematical friends approached the Guinness company to help pay the costs of printing his collected papers in a single volume. Whether this is true or not, it is clear from the memoirs of the American statistician Harold Hotelling, who in the late 1930s wanted to talk with "Student," that arrangements were made to meet him secretly, with all the aspects of a spy mystery. This suggests that the true identity of "Student" was still a secret from the Guinness company. The papers by "Student" that were printed in *Biometrika* lay on the cusp of theory and application, as Gosset moved from highly practical problems into difficult mathematical formulations and back out into practical reality with solutions that others would follow.

In spite of his great achievements, Gosset was an unassuming man. In his letters, one frequently finds expressions like "my own investigations [provide] only a rough idea of the thing....," or protestations that he was given too much credit for some discovery when "Fisher really worked out the complete mathematics...." Gosset, the man, is remembered as a kind and thoughtful colleague who was sensitive to the emotional problems of others. When he died at the age of sixty-one, he left his wife, Marjory (who was a vigorous athlete, having been captain of the English Ladies Hockey Team), one son, two daughters, and a grandson. His parents were still alive at the time of his death.

"STUDENT"'S T-TEST

If nothing else, all scientists are indebted to Gosset for a short paper entitled "The Probable Error of the Mean," which appeared in

Biometrika in 1908. It was Ronald Aylmer Fisher who pointed out the general implications of this remarkable paper. To Gosset, there was a specific problem to be solved, which he attacked in the evenings in his home with his usual patience and carefulness. Having discovered a solution, he checked it against other data, reexamined his results, tried to determine if he had missed any subtle differences, considered what assumptions he had to make, and calculated and recalculated his findings. He anticipated modern computer Monte Carlo techniques, where a mathematical model is simulated over and over again to determine the probability distributions associated with it. However, he had no computer. He painstakingly added up numbers, taking averages from hundreds of examples and plotting the resulting frequencies—all by hand.

The specific problem Gosset attacked dealt with small samples. Karl Pearson had computed the four parameters of a distribution by accumulating thousands of measurements from a single distribution. He assumed that the resulting estimates of the parameters were correct because of the large samples he used. Fisher was to prove him wrong. In Gosset's experience, the scientist seldom had the luxury of such large samples. More typical was an experiment yielding ten to twenty observations. Furthermore, he recognized this as very common in all of science. In one of his letters to Pearson, he writes: "If I am the only person that you've come across that works with too small samples, you are very singular. It was on this subject that I came to have dealings with Stratton [a fellow at Cambridge University, where] ... he had taken as an illustration a sample of 4!"

All of Pearson's work assumed that the sample of data was so large that the parameters could be determined without error. Gosset asked, what happens if the sample were small? How can we deal with the random error that is bound to find its way into our calculations?

Gosset sat at his kitchen table at night, taking small sets of numbers, finding the average and the estimated standard deviation, dividing one by the other, and plotting the results on graph paper.

He found the four parameters associated with this ratio and matched it to one of Pearson's family of skew distributions. His great discovery was that you did not have to know the exact values of all four parameters of the original distribution. The ratios of the estimated values of the first two parameters had a probability distribution that could be tabulated. It did not matter where the data came from or what the true value of the standard deviation was. Taking the ratio of these two sample estimates brought you into a known distribution.

As Frederick Mosteller and John Tukey point out, without this discovery, statistical analysis was doomed to use an infinite regression of procedures. Without "Student"'s t,[2] as the discovery has come to be called, the analyst would have to estimate the four parameters of the observed data, then estimate the four parameters of the estimates of the four parameters, then estimate the four parameters of each of those, and so on, with no chance of ever reaching a final calculation. Gosset showed that the analyst could stop at the first step.

There was a fundamental assumption in Gosset's work. He assumed that the initial set of measurements had a normal distribution. Over the years, as scientists used "Student"'s t, many came to believe that this assumption was not needed. They often found that "Student"'s t had the same distribution regardless of whether the initial measurements were normally distributed or not. In 1967, Bradley Efron of Stanford University proved that this was true. To be more exact, he found the general conditions under which the assumption was not needed.

With the development of "Student"'s t, we are sliding into a use of statistical distribution theory that is widespread in the sciences but with which there are deep philosophical problems. This is the

[2]This is an example of what might be considered a corollary of Stigler's law of misonomy. Gosset used the letter "z" to indicate this ratio. Some years later, writers of textbooks developed the tradition of referring to normally distributed variables with the letter "z," and they began using the letter "t" for "Student"'s ratio.

use of what are called "hypothesis tests," or "significance tests." We shall explore this in a later chapter. For the moment, we only note that "Student" provided a scientific tool that almost everyone uses—even if few really understand it.

In the meantime, "dear Mr. Gosset" became the middleman between two towering and feuding geniuses, Karl Pearson and Ronald Aylmer Fisher. He maintained a close friendship with both, although he often complained to Pearson that he did not understand what Fisher had written him. His friendship with Fisher began while the latter was still an undergraduate at Cambridge University. Fisher's tutor[3] in astronomy introduced them in 1912, when Fisher had just become a wrangler (the highest mathematical honor) at Cambridge. He was looking at a problem in astronomy and wrote a paper in which he rediscovered "Student"'s 1908 results—young Fisher was obviously unaware of Gosset's previous work.

In this paper, which Fisher showed Gosset, there was a small error, which Gosset caught. When he returned home, he found waiting for him two pages of detailed mathematics from Fisher. The young man had redone the original work, expanded upon it, and identified an error that Gosset had made. Gosset wrote to Pearson: "I am enclosing a letter which gives a proof of my formulae for the frequency distribution of ["Student"'s t].... Would you mind looking at it for me; I don't feel at home in more than three dimensions, even if I could understand it otherwise...." Fisher had proved Gosset's results using multidimensional geometry.

In the letter, Gosset explains how he went to Cambridge to meet with a friend who was also Fisher's tutor at Gonville and Caius College and how he was introduced to the 22-year-old student. He goes on: "This chap Fisher produced a paper giving 'a

[3]It is the practice at British universities like Cambridge for each student to have a faculty member, called the student's tutor, who guides the student through the appropriate course work.

new criterion of probability' or something of the sort. A neat but as far as I could understand it, quite unpractical and unserviceable way of looking at things."

After describing his discussion with Fisher at Cambridge, Gosset writes:

> To this he replied with two foolscap pages covered with mathematics of the deepest dye in which he proved [this is followed by a group of mathematical formulas].... I couldn't understand the stuff and wrote and said I was going to study it when I had time. I actually took it up to the Lakes with me—and lost it!
>
> Now he sends this to me. It seemed to me that if it's all right perhaps you might like to put the proof in a note. It's so nice and mathematical that it might appeal to some people....

Thus did one of the great geniuses of the twentieth century burst upon the scene. Pearson published the young man's note in *Biometrika.* Three years later, after a sequence of very condescending letters, Pearson published a second paper by Fisher, but only after making sure that it would be seen as a minor addition to work done by one of Pearson's coworkers. Pearson never allowed another paper by Fisher into his journal. Fisher went on to find errors in many of Pearson's proudest accomplishments, while Pearson's editorials in later issues of *Biometrika* often refer to errors made by "Mr. Fisher" or by "a student of Mr. Fisher" in papers in other journals. All of this is grist for the next chapter. Gosset will appear again in some of the succeeding chapters. As a genial mentor, he was instrumental in introducing younger men and women to the new world of statistical distribution, and many of his students and coworkers were responsible for major contributions to the new mathematics. Gosset, in spite of his modest protestations, made many long-lasting contributions of his own.

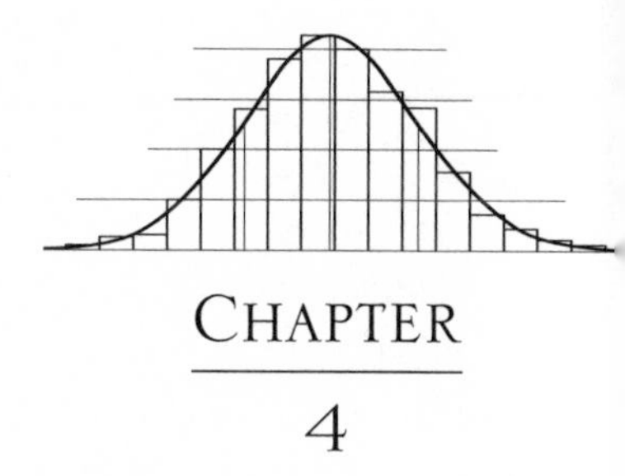

CHAPTER

4

RAKING OVER THE MUCK HEAP

Ronald Aylmer Fisher was twenty-nine years old when he moved with his wife, three children, and sister-in-law into an old farmhouse near the Rothamsted Agricultural Experimental Station, north of London, in the spring of 1919. By many measures, he could have been considered a failure in life. He had grown up as a sickly and lonely child with severe vision impairment. To protect his nearsighted eyes, the doctors had forbidden him to read by artificial light. He had taken early to mathematics and astronomy. He was fascinated with astronomy at age six. At ages seven and eight, he was attending popular lectures given by the famous astronomer Sir Robert Ball.

Fisher matriculated at Harrow, the renowned public school,[1] where he excelled in mathematics. Because he was not allowed to use electric light, his mathematics tutor would teach him in the evening without the use of pencil, paper, or any other visual aids. As a result, Fisher developed

[1]Misonomy extends beyond mathematics. In England, the most exclusive private secondary schools, like Harrow, are called "public schools."

a deep geometric sense. In future years, his unusual geometric insights enabled him to solve many difficult problems in mathematical statistics. The insights were so obvious to him that he often failed to make them understandable to others. Other mathematicians would spend months or even years trying to prove something that Fisher claimed was obvious.

He entered Cambridge in 1909, rising to the prestigious title of wrangler in 1912. A student at Cambridge becomes a wrangler by passing a series of extremely difficult mathematics exams, both oral and written. This was something accomplished by no more than one or two classmen a year, and some years there were no wranglers. While still an undergraduate, Fisher published his first scientific paper, where complicated iterative formulas are interpreted in terms of multidimensional geometric space. In this paper, what had hitherto been an exceedingly complicated method of computation is shown to be a simple consequence of that geometry. He stayed for a year after graduation to study statistical mechanics and quantum theory. By 1913, the statistical revolution had entered physics, and these were two areas where the new ideas were sufficiently well formulated to produce formal course work.

Fisher's first job was in the statistical office of an investment company, which he left suddenly to do farmwork in Canada; again, he left his work suddenly to return to England at the beginning of the First World War. Although he qualified for a commission in the army, his poor eyesight kept him out of military service. He spent the war years teaching mathematics in a series of public schools, each experience worse than the previous one. He was short-tempered with students who could not understand what was to him obvious.

FISHER VERSUS KARL PEARSON

While still an undergraduate, Fisher had a note published in *Biometrika*, as mentioned in the previous chapter. As a result, Fisher met Karl Pearson, who introduced him to the difficult problem of determining the statistical distribution of Galton's correlation coefficient. Fisher thought about the problem, cast it into a geometric formulation, and within a week had a complete answer. He submitted it to Pearson for publication in *Biometrika;* Pearson could not understand the mathematics and sent the paper to William Sealy Gosset, who also had difficulty understanding it. Pearson knew how to get partial solutions to the problem for specific cases. His method involved monumental amounts of calculation, and he set the workers in his biometrical laboratory to computing those specific answers. In every case, they agreed with Fisher's more general solution. Still, Pearson did not publish Fisher's paper. He urged Fisher to make changes and to reduce the generality of the work. Pearson held Fisher off for over a year, while he had his assistants (the "calculators") computing a large, extensive table of the distribution for selected values of the parameters. Finally, he published Fisher's work, but as a footnote to a larger paper in which he and one of his assistants displayed these tables. The result was that, to the casual reader, Fisher's mathematical manipulations were a mere appendix to the more important and massive computational work done by Pearson and his coworkers.

Fisher never published another paper in *Biometrika*, although this was the preeminent journal in the field. In the following years, his papers appeared in the *Journal of Agricultural Science*, *The Quarterly Journal of the Royal Meteorological Society*, *The Proceedings of the Royal Society of Edinburgh*, and the *Proceedings of the Society of Psychical Research*. All of these are journals that one does not normally associate with mathematical research. According to some who knew Fisher, these choices were made because Pearson and his friends effectively froze Fisher out of the mainstream of mathematical and statistical research. According to others, Fisher himself felt rebuffed by Pearson's cavalier attitude and by his failure to get a similar paper published in the *Journal of the Royal Statistical Society* (the other prestigious journal in the field);

and he proceeded to use other journals, sometimes paying the journal to have his paper appear in it.

FISHER THE "FASCIST"

Some of these early papers by R. A. Fisher are highly mathematical. The paper on the correlation coefficient, which Pearson finally published, is dense with mathematical notation. A typical page is half or more filled with mathematical formulas. There are also papers in which no mathematics appear at all. In one of them, he discusses the ways in which Darwin's theory of random adaptation is adequate to account for the most sophisticated anatomical structures. In another, he speculates on the evolution of sexual preference. He joined the eugenics movement and, in 1917, published an editorial in the *Eugenics Review*, calling for a concerted national policy "to increase the birth-rate in the professional classes and among highly skilled artisans" and to discourage births among the lower classes. He argued in this paper that governmental policies that provided welfare for the poor encouraged them to procreate and pass on their genes to the next generation, whereas the concerns of the middle class for economic security led to postponement of marriages and limited families. The end result, Fisher feared, was for the nation to select the "poorest" genes for future generations and to deselect the "better" genes. The question of eugenics, the movement to improve the human gene stock by selective breeding, would dominate much of Fisher's political views. During World War II, he would be falsely accused of being a fascist and squeezed out of all war-related work.

Fisher's politics contrast with the political views of Karl Pearson, who flirted with socialism and Marxism, whose sympathies lay with the downtrodden, and who loved to challenge the entrenched "better" classes. Whereas Pearson's political views had little obvious effect on his scientific work, Fisher's concern over eugenics led him to put a great deal of effort into the mathematics of genetics.

Starting with the (at that time) new ideas that specific characteristics of a plant or animal can be attributed to a single gene, which can occur in one of two forms, Fisher moved well beyond the work of Gregor Mendel,[2] showing how to estimate the effects of neighboring genes on each other.

The idea that there are genes that govern the nature of life is part of the general statistical revolution of science. We observe characteristics of plants and animals that are called "phenotypes," but we postulate that these phenotypes are the result of interactions among the genes with different probabilities of interaction. We seek to describe the distribution of phenotypes in terms of these underlying and unseen genes. In the late twentieth century, biologists identified the physical nature of these genes as segments of the hereditary molecule, DNA. We can read these genes to determine what proteins they instruct the cell to make, and we talk about these as real events. But what we observe is still a scatter of possibilities, and the segments of DNA we call genes are imputed from that scatter.

This book deals with the general statistical revolution, and R. A. Fisher played an important role in it. He was proud of his achievements as a geneticist, and about half of his output deals with genetics. We will leave Fisher, the geneticist, at this point and look at Fisher primarily in terms of his development of general statistical techniques and ideas. The germs of these ideas can be found in his

[2]Gregor Mendel was a Central European monk (whose real first name was Johann—more misonomy) who in the 1860s published a set of articles describing experiments on the breeding of peas. His work fell into obscurity, since it did not fit the general pattern of botanical work then being published. It was rediscovered by a group of biologists at Cambridge University under the leadership of William Bateson, who established a department of genetics at Cambridge. One of the many controversies that Karl Pearson seemed to enjoy consisted of his disdain for the work of these geneticists, who examined minute discrete changes in living organisms, whereas Pearson was interested in grand continuous modification of parameters as the true nature of evolution. One of Fisher's first papers showed that Pearson's formulas could be derived from Bateson's minute discrete changes. Pearson's comments on seeing this were that it was obvious and that Fisher should send the paper to Bateson to show Bateson the truth. Bateson's comments were that Fisher should send it to Pearson to show Pearson the truth. Eventually, Fisher succeeded Bateson as chairman of the Department of Genetics at Cambridge.

early papers but were more fully developed as he worked at Rothamsted during the 1920s and early 1930s.

STATISTICAL METHODS FOR RESEARCH WORKERS

Although Fisher was ignored by the mathematical community during that time, he published papers and books that greatly influenced the scientists working in agriculture and biology. In 1925, he published the first edition of *Statistical Methods for Research Workers*. This book went through fourteen English-language editions and appeared in French, German, Italian, Japanese, Spanish, and Russian translations.

Statistical Methods for Research Workers was like no other mathematics book that had appeared before it. Usually, a book of mathematics has theorems and proofs of those theorems, and develops abstract ideas and generalizes them, relating them to other abstract ideas. If there are applications in such books, they occur only after the mathematics have been fully described and proven. *Statistical Methods for Research Workers* begins with a discussion of how to create a graph from numbers and how to interpret that graph. The first example, occurring on the third page, displays the weight of a baby each week for the first thirteen weeks of life. That baby was Fisher's firstborn, his son George. The succeeding chapters describe how to analyze data, giving formulas, showing examples, interpreting the results of those examples, and moving on to other formulas. None of the formulas is derived mathematically. They all appear without justification or proof. They are often presented with detailed techniques of how to implement them on a mechanical calculator, but no proofs are displayed.

Despite, or perhaps because of, its lack of theoretical mathematics, the book was rapidly taken up by the scientific community. It met a serious need. It could be handed to a lab technician with minimum mathematical training, and that technician could use it. The scientists who used it took Fisher's assertions as correct. The mathematicians who reviewed it looked askance at its audacious unproven statements, and many wondered how he had come to these conclusions.

During the Second World War, the Swedish mathematician Harald Cramér, isolated by the war from the international scientific community, spent days and weeks reviewing this book and Fisher's published papers, filling in the missing steps of proofs and deriving proofs where none were indicated. In 1945, Cramér produced a book entitled *Mathematical Methods of Statistics*, giving formal proofs for much of what Fisher had written. Cramér had to choose among the many outpourings of this fertile genius, and a great deal that Fisher had written was not included in this book. Cramér's book was used to teach a generation of new mathematicians and statisticians, and his redaction of Fisher became the standard paradigm. In the 1970s, L. J. Savage of Yale University went back to Fisher's original papers and discovered how much Cramér had left out. He was amazed to see that Fisher had anticipated later work by others and had solved problems that were thought to be still unsolved in the 1970s.

But all of this was still in the future in 1919, when Fisher abandoned his unsuccessful career as a schoolmaster. He had just finished a monumental work, combining Galton's correlation coefficient and the gene theory of Mendelian heredity. The paper had been rejected by the Royal Statistical Society and by Pearson at *Biometrika*. Fisher heard that the Royal Society of Edinburgh was looking for papers to publish in its *Transactions*, but that the authors were expected to pay for the publication costs. Thus, he paid to have his next great mathematical work published in an obscure journal.

At this point, Karl Pearson, still impressed by young Fisher, came through with an offer to take him on as chief statistician at the Galton Biometrical Laboratory. The correspondence between the two men was cordial, but it was obvious to Fisher that Pearson was strong-willed and dominating. His chief statistician would, at best, be engaged in detailed calculations that were dictated by Pearson.

ROTHAMSTED AND AGRICULTURAL EXPERIMENTS

Fisher had also been contacted by Sir John Russell, head of the Rothamsted Agricultural Experimental Station. The Rothamsted station had been set up by a British maker of fertilizer on an old farm that had once belonged to the original owners of the fertilizer firm. The clay soil was not particularly suited to growing much of anything, but the owners had discovered how to combine crushed stone with acid to produce what was known as Super-Phosphate. The profits from the production of Super-Phosphate were used to establish an experimental station where new artificial fertilizers might be developed. For ninety years, the station ran "experiments," testing different combinations of mineral salts and different strains of wheat, rye, barley, and potatoes. This had created a huge storehouse of data, exact daily records of rainfall and temperature, weekly records of fertilizer dressings and measures of the soil, and annual records of harvests—all of it preserved in leather-bound notebooks. Most of these experiments had not produced consistent results, but the notebooks had been carefully stored away in the station's archives.

Sir John looked at this vast collection of data and decided that maybe somebody might be hired to see what was in it, take a sort of statistical look at these records. He inquired around, and someone recommended Ronald Aylmer Fisher. He offered Fisher a year's employment at a thousand pounds; he could not offer more and could not guarantee that the job would last past that one year.

Fisher accepted Russell's offer. He took his wife, his sister-in-law, and three children into the rural area north of London. They rented a farm next door to the experimental station, where his wife and sister-in-law tended a vegetable garden and kept house for him. He put on his boots and walked across the fields to Rothamsted Agricultural Experimental Station and its ninety years of data, to engage in what he was later to call "raking over the muck heap."

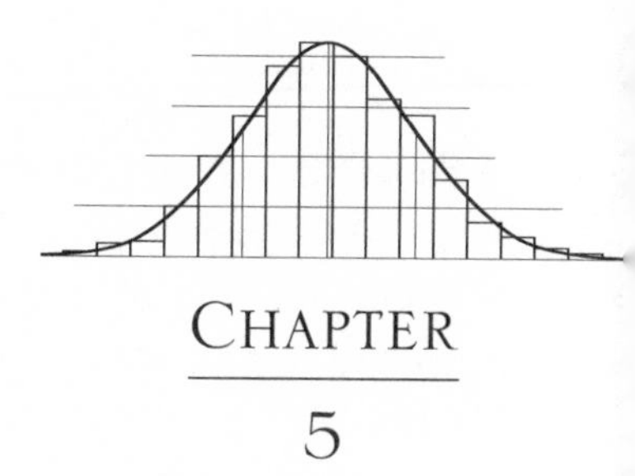

CHAPTER

5

"STUDIES IN CROP VARIATION"

Early in my career as a biostatistician, and on one of my trips to the University of Connecticut at Storrs to talk over my problems, Professor Hugh Smith had a present for me. It was a copy of a paper entitled "Studies in Crop Variation. III. The Influence of Rainfall on the Yield of Wheat at Rothamsted." The paper was fifty-three pages long. It is the third in a series of remarkable mathematical articles, the first of which had appeared in the *Journal of Agricultural Science*, vol. XI, in 1921. Variation in output is the bane of the experimental scientist but the basic material of statistical methods. The word *variation* is seldom used in the modern scientific literature. It has been replaced by other terms, like "variance," which refer to specific parameters of distributions. Variation is too vague a word for ordinary scientific use, but it was appropriate in this series of papers, because the author used the variation in crop outputs from year to year and from field to field as a starting point from which to derive new methods of analysis.

Most scientific papers have long lists of references at the end, identifying the previous papers that had addressed the problems discussed. "Studies in Crop Variation. I," the first in this series, had only three references, one indicating a failed attempt made in 1907 to correlate rainfall and wheat growth; a second, in German, from 1909, describing a method for computing the minimum of a complicated mathematical formula; and the third, a set of tables published by Karl Pearson. There was no previous paper dealing with most of the topics covered in this remarkable series. The "Studies in Crop Variation" were sui generis. Their author is described as R. A. Fisher, M.A., Statistical Laboratory, Rothamsted Experimental Station, Harpenden.

In 1950, the publisher John Wiley asked Fisher if he would cull through his published papers and provide a selection of the most important, to form a single volume. The volume, entitled *Contributions to Mathematical Statistics*, opens with a contemporaneous photograph of a white-haired R. A. Fisher, his lips tightly pressed together, his tie slightly askew, his white beard not too well trimmed. He is identified as "R. A. Fisher, Department of Genetics, University of Cambridge." "Studies in Crop Variation. I" is the third article in this book. It is preceded by a short note from the author identifying its importance and its place in his works:

> In the author's early work at Rothamsted much attention was given to the massive records of weather, crop yields, crop analyses, etc., which had been accumulated during the long history of that research station. The material was obviously of unique value for such problems as that of ascertaining to what extent meteorological readings were capable of supplying a prediction of the crop yields to follow. The present paper is the first of a series devoted to this end. . . .

There were, at most, six articles in the "series devoted to this end." "Studies in Crop Variation. II" appeared in 1923. There is the paper Professor Smith gave me, entitled "III. The Influence of Rainfall on the Yield of Wheat at Rothamsted," from 1924. "Studies in Crop Variation. IV" appeared in 1927. "Studies in Crop Variation. VI" was published in 1929. Study number V does not appear in Fisher's collected works. Seldom in the history of science has a set of titles been such a poor descriptor of the importance of the material they contain. In these papers, Fisher developed original tools for the analysis of data, derived the mathematical foundations of those tools, described their extensions into other fields, and applied them to the "muck" he found at Rothamsted. These papers show a brilliant originality and are filled with fascinating implications that kept theoreticians busy for the rest of the twentieth century, and will probably continue to inspire more work in the years that follow.

"Studies in Crop Variation. I"

There were additional authors for two of the later papers in Fisher's series. For "Studies in Crop Variation. I," he worked alone. It required prodigious amounts of calculation. His only aid was a calculating machine named the Millionaire. It was a primitive, hand-cranked mechanical calculator. If one wanted to multiply, for instance, 3,342 by 27, one put the platen on the units position, set in the number 3,342, and cranked seven times. Then one put the platen on the tens position, set in the number 3,342, and cranked two times. It was a Millionaire because the platen was big enough to accommodate numbers in the millions.

To get some idea of the physical effort involved, consider Table VII that appears on page 123 of "Studies in Crop Variation. I." If it took about one minute to complete a single large-digit multiplication, I estimate that Fisher needed about 185 hours of work to generate that table. There are fifteen tables of similar complexity and four large complicated graphs in the article. In terms of physical labor alone, it must have taken at least eight months of 12-hour days to prepare the tables for this article. This does not include the hours needed to work out the theoretical mathematics, to organize the data, to plan the analysis, and to correct the inevitable mistakes.

GALTON'S REGRESSION TO THE MEAN GENERALIZED

Recall Galton's discovery of regression to the mean and his attempt to find a mathematical formula that linked random events to each other. Fisher took Galton's word, *regression*, and established a general mathematical relationship between the year and the wheat output of a given field. Pearson's idea of probability distribution now became a formula connecting year to output. The parameters of this more complicated distribution described different aspects of the change in wheat output. To go through Fisher's mathematics one needs a solid knowledge of calculus, a good sense of the theory of probability distributions, and a feeling for multidimensional geometry. But it is not too difficult to understand his conclusions.

He pulled the time trend of wheat output into several pieces. One was a steady overall diminution of output due to deterioration of the soil. Another was a long-term, slow change that took several years for each phase. The third was a set of faster moving changes that took into account the variations in climate from year to year. Since Fisher's first pioneering attempts, the statistical analysis of time series has built on his ideas and methods. We now have computers that can do the immense calculations with clever algorithms, but the basic idea and methods remain. Given a set of numbers fluctuating over time, we can pull them apart into effects that are due to different sources. Time series analyses have been used to examine the lapping of waves on the Pacific shores of the United States and thereby identify storms in the Indian Ocean. These methods have enabled researchers to distinguish between underground nuclear explosions and earthquakes, pinpoint pathological aspects of heartbeats, quantify the effect of environmental regulations on air quality; and the uses continue to multiply.

Fisher was puzzled by his analysis of grain harvested from a field named Broadbalk. Only natural animal dung had been used on this field, so the variation in yield from year to year was not the result of experimental fertilizers. The long-term deterioration made sense as the soil was depleted of nutrients missing from the dung, and he could identify the effects of differing rainfall patterns in the year-to-year changes. What was the source of the slow changes? The slow-change pattern suggested that in 1876, the output had begun to deteriorate more than would be expected from the other two effects, the deterioration getting more rapid after 1880. There was an improvement starting in 1894 and continuing until 1901, with a drop thereafter.

Fisher found another record with the same slow change but with the pattern reversed. This was the infestation of weeds in the wheat field. After 1876, the weeds became ever heavier, with new varieties of perennials establishing themselves. Then in 1894 the weeds suddenly began to diminish, only to start flourishing again in 1901.

It turned out that it had been the practice prior to 1876 to hire small boys to go into the fields and pull weeds. It was common at that time to see weary children in the fields of England on an afternoon combing through the wheat and other grains, constantly pulling weeds. In 1876, the Education Act made attendance at school compulsory, and the legions of young boys began disappearing from the fields. In 1880, a second Education Act provided penalties for families that kept their children out of school, and the last of the young boys left the fields. Without the little fingers to pull them out, the weeds began to flourish.

What happened in 1894 to reverse this trend? There was a boarding school for girls in the vicinity of Rothamsted. The new schoolmaster, Sir John Lawes, believed in vigorous outdoor activity to build up the health of his young charges. He arranged with the director of the experimental station to bring his young girls out into the fields to pull weeds on Saturdays and evenings. After Sir John died in 1901, the little girls went back to sedentary and indoor activities, and the weeds came back to Broadbalk.

RANDOMIZED CONTROLLED EXPERIMENTS

The second study on crop variation also appeared in the *Journal of Agricultural Science* in 1923. This paper does not deal with accumulated data from the past experiments at Rothamsted. Instead, it describes a set of experiments on the effects of different fertilizer mixes on different varieties of potatoes. Something remarkable had happened to the experiments at Rothamsted since the arrival of R. A. Fisher. No longer were they applying a single experimental fertilizer across an entire field. Now, they were cutting the field up into small plots. Each plot was further divided into rows of plants and each row within a plot was given a different treatment.

The basic idea was a simple one—simple, that is, once it was proposed by Fisher. No one had ever thought of it before. It is obvious to anyone looking at a field of grain that some parts of the field are better than others. In some corners, the plants grow tall and heavy with grain. In other corners, the plants are thin and scraggly. It may be due to the way in which water drains, or to the changes in soil type, or the presence of unknown nutrients, or blocks of perennial weeds, or some other, unforeseen, effect. If the agricultural scientist wants to test the difference between two fertilizer components, he can put one fertilizer on one part of the field and another elsewhere. This will confound the effects of the fertilizer with effects due to the soil or drainage properties. If the tests are run on the same fields but in different years, the effects of fertilizer are confounded with changes in weather from year to year.

If the fertilizers are compared right next to one another and in the same year, then the differences in soil will be minimized. They will still be there, since the treated plants are not on exactly the same soil. If we use many such pairs, the differences due to soil will average out in some sense. Suppose we want to compare two fertilizers, one with twice as much phosphorus as the other. We break the field into tiny plots, each with two rows of plants. We always put the extra phosphorus on the northern row of plants and treat the southern row with the other mixture. Now, I can hear someone saying, then they won't "average" out if the fertility gradient in the soil runs north and south, for the northern row in each block will have slightly better soil than the southern row.

We'll just alternate. In the first block, the extra phosphorus will be in the northern row. In the second block, it will be in the southern row, and so forth. One of my readers has now drawn a rough map of the field and put X's to indicate the rows with extra phosphorus. He points out that if the fertility gradient runs from northwest to southeast, then the rows treated with extra phosphorus will all have better soil than the others. Someone else points out that if the gradient runs from northeast to southwest, the opposite holds. Well, demands another reader, which is it? How does the fertility gradient run? To which we reply, no one knows. The concept of a fertility gradient is an abstract one. The real pattern of fertility may run up and down in some complicated way as we go from north to south or east to west.

I can imagine these discussions among the scientists at Rothamsted once Fisher pointed out that setting treatments within small blocks would allow for more careful experimentation. I can imagine the discussions over how to determine the fertility gradient, while R. A. Fisher sits back and smiles, letting them get more and more involved in complicated constructs. He has already considered these questions and has a simple answer. He removes the pipe from his mouth. Those who knew him, describe Fisher as sitting, quietly puffing on his pipe, while arguments raged about him, waiting for the moment when he could insert his answer. "Randomize," he says.

FISHER'S ANALYSIS OF VARIANCE

It is simple. The scientist assigns the treatments to different rows within a block at random. Since the random ordering follows no fixed pattern, any possible structure of fertility gradient will cancel out, on the average. Fisher springs up and begins to write furiously on the blackboard, filling it with mathematical symbols, sweeping his arms across the columns of math, crossing out factors that cancel on either side of the equations, and emerging with what was to become probably the single most important tool of biological science. This is a method for separating out the effects of different treatments in a well-designed scientific experiment, which Fisher called the "analysis of variance." In "Studies in Crop Variation. II" the analysis of variance shows up for the first time.

The formulas for some examples of analysis of variance appear in *Statistical Methods for Research Workers*, but in this paper they are derived mathematically. They are not worked out in sufficient detail for an academic mathematician to be satisfied. The algebra displayed is made specific to the situation of comparing three types of fertilizer (manure), ten varieties of potatoes, and four blocks of soil. It takes a few hours of painstaking work to figure out how the algebra might be adapted for two fertilizers and five varieties, or for six fertilizers and only one variety. It takes even more mathematical sweat of the brow to figure out the general formulas that would work in all cases. Fisher, of course, knew the general formulas. They were so obvious to him that he did not see the need to produce them.

No wonder his contemporaries were mystified by young Fisher's work!

"Studies in Crop Variation. IV" introduces what Fisher called "analysis of covariance." This is a method for factoring out the effects of conditions that are not part of the experimental design but which are there and can be measured. When an article in a medical journal describes a treatment effect that has been "adjusted for sex and weight," it is using the methods pioneered by Fisher in this paper. Study VI produces refinements in the theory of design of experiments. Study III, to which Professor Smith introduced me, will be discussed later in this chapter.

DEGREES OF FREEDOM

In 1922, Fisher finally got his first article published in the *Journal of the Royal Statistical Society*. It is a short note that modestly proves that one of Karl Pearson's formulas was wrong. Writing about this paper many years later, Fisher said:

> This short paper, with all its juvenile inadequacies, yet did something to break the ice. Any reader who feels exasperated by its tentative and piecemeal character should remember that it had to find its way to publication past critics who, in the first place, could not believe that Pearson's work stood in need of correction, and who, if this had to be admitted, were sure that they themselves had corrected it.

In 1924, he was able to publish a longer and more general paper in the *Journal of the Royal Statistical Society*. He later comments on this and a related paper in an economics journal: "[These papers] are attempts to reconcile, with the aid of the new concept of degrees of freedom, the discrepant and anomalous results observed by different authors. . . ."

The "new concept of degrees of freedom" was Fisher's discovery and was directly related to his geometric insights and his ability to cast the mathematical problems in terms of multidimensional geometry. The "anomalous results" occurred in an obscure book published by somebody named T. L. Kelley in New York; Kelley had found data wherein some of Pearson's formulas did not seem to produce correct answers. Only Fisher, it would seem, had looked at Kelley's book. Kelley's anomalous results were only used as a springboard from which Fisher utterly demolished another of Pearson's most proud achievements.

"STUDIES IN CROP VARIATION. III"

The third of the studies in crop variation appeared in 1924 in the *Philosophical Transactions of the Royal Society of London*. It begins:

> At the present time very little can be claimed to be known as to the effects of weather upon farm crops. The obscurity of the subject, in spite of its immense importance to a great national industry, may be ascribed partly to the inherent complexity of the problem . . . and . . . to the lack of quantitative data obtained either upon experimental or under industrial conditions. . . .

Then follows a masterful article of fifty-three pages. In it lie the foundations of modern statistical methods used in economics, medicine, chemistry, computer science, sociology, astronomy, pharmacology—any field where one needs to establish the relative effects of a large number of interconnected causes. It contains highly ingenious methods of calculation (recall that Fisher had only the manual Millionaire to work with), and many wise suggestions on how to organize data for statistical analysis. I am forever grateful to Professor Smith for introducing me to this paper. With each reading, I learn something new.

The first volume (of five) of the *Collected Papers of R. A. Fisher* ends with the papers he published in 1924. There is a photograph of Fisher, at this point thirty-four years old, near the end of the volume. His arms are folded. His beard is well trimmed. His glasses do not seem as thick as they did in earlier photos. He has a confident and secure look. In the previous five years, he has built a remarkable statistics department at Rothamsted. He has hired coworkers like Frank Yates, who would go on, under Fisher's encouragement, to make major contributions to the theory and practice of statistical analysis. With a few exceptions, Karl Pearson's students have disappeared. While they worked at the biometrical laboratory, they aided Pearson and were no more than extensions of Pearson. With few exceptions, Fisher's students responded to his encouragement and plowed brilliant and original paths themselves.

In 1947, Fisher was invited to give a series of talks on the BBC radio network about the nature of science and scientific investigation. Early in one talk, he said:

> A scientific career is peculiar in some ways. Its raison d'être is the increase of natural knowledge. Occasionally, therefore, an increase of natural knowledge occurs. But this is tactless, and feelings are hurt. For in some small degree it is inevitable that views previously expounded are shown to be either obsolete or false. Most people, I think, can recognize this and take it in good part if what they have been teaching for ten years or so comes to need a little revision; but some undoubtedly take it hard, as a blow to their amour propre, or even as an invasion of the territory they have come to think of as exclusively their own, and they must react with the same ferocity as we can see in the robins and chaffinches these spring days when they resent an intrusion into their little territories. I do not think anything can be done about it. It is inherent in the nature of our profession; but a young scientist may be warned and advised that when he has a jewel to offer for the enrichment of mankind some certainly will wish to turn and rend him.

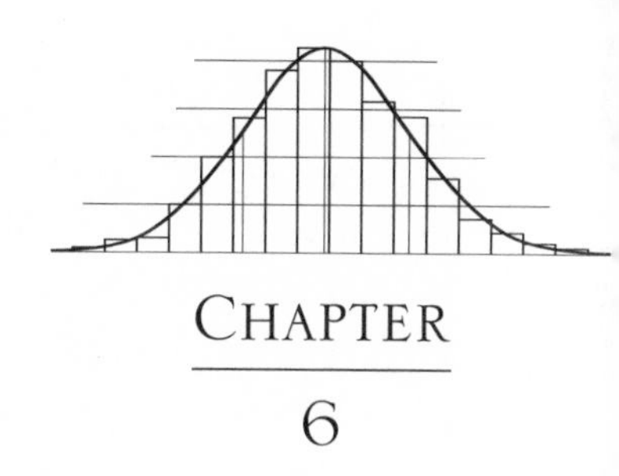

CHAPTER

6

"THE HUNDRED-YEAR FLOOD"

What can be more unpredictable than the "hundred-year flood," the flood waters that come pouring down a river with a massive ferocity so great that it happens as rarely as once in a hundred years? Who can plan for such an event? How can we estimate the height of flood waters that come so infrequently? If the statistical models of modern science deal with the distribution of many observations, what can statistical models do for the problem of the flood that has never been seen or, if it has occurred, been seen only once? L. H. C. Tippett found a solution.

Leonard Henry Caleb Tippett was born in 1902 in London and studied physics there at Imperial College, graduating in 1923. He was attracted, he said, to physics because of its "insistence on accurate measurement . . . and a disciplined approach to the scientific controversies of the day." Looking back at his youthful enthusiasm, he continued: "We tended to think of a hypothesis as being either right or wrong, and regarded the crucial experiment as being the main instrument for advancing knowledge." When he had the opportunity to do experiments, he found that the experimental results never agreed exactly with what the theory would predict. By his own experience, he said, "I found it better to improve the sampling technique [here he refers to statistical distributions] than to discard the theory." Tippett realized that the theory he so loved provided information only about parameters and not about individual observations.

In this way, L. H. C. Tippett (as he is identified in his published papers) became attuned to the statistical revolution through his own understanding of experimentation. Upon graduation, he took a position as a statistician at the British Cotton Industry Research Association, usually called the Shirley Institute, where attempts were being made to improve the manufacture of cotton thread and cloth through the use of modern scientific methods. One of the most vexing problems dealt with the strength of newly spun cotton thread. The tension needed to break a strand of thread differed greatly from one strand to another, even when the strands had been spun under identical circumstances. Tippett ran some careful experiments, examining the thread under the microscope after different levels of tension. He discovered that the breaking of the thread depended upon the strength of the weakest fiber in it.

The weakest fiber? How would one model the mathematics of the strength of the weakest fiber? Unable to solve this problem, Tippett asked for and was given a year's leave of absence in 1924 to study under Karl Pearson at the Galton Biometrical Laboratory at University College, London. Of that experience, Tippett wrote:

> The time at University College was thrilling. Karl Pearson was a great man, and we felt his greatness. He was hard-working and enthusiastic and inspired his staff and students. When I was there he was still doing research, and would come into his lectures, full of excitement and enthusiasm, giving results hot from his study table. That in those years his lines of research were somewhat out of date did not make his lectures any

> less stimulating. . . . It was typical of his breadth of interests that one of his lecture courses was on "The History of Statistics in the 17th and 18th Centuries.". . . He was a vigorous controversialist . . . and one series of publications issued by him is termed "Questions of the Day and of the Fray" . . . the influence of the vigorous and controversial past was in the atmosphere. The walls of the department were embellished with mottoes and cartoons . . . there was . . . a cartoon by "Spy" which was a caricature of "Soapy Sam"—the Bishop Wilberforce who had the famous verbal duel with T. H. Huxley on Darwinism at the British Association meeting in 1860. There was a display of publications that had been issued over past decades, and an impression of the interests of the department was given by such titles as "Treasury of Human Inheritance (Pedigrees of Physical, Psychical, and Pathological Characters in Man)" and "Darwinism, Medical Progress and Eugenics." K. P. reminded us of his close connection with Galton at an annual departmental dinner when he gave a description of the year's work in the form of a report he would have given Galton, had he been alive. And we toasted "the biometric dead."

This was Karl Pearson in the final active years of his life, before the work of R. A. Fisher and Pearson's own son was to sweep most of his scientific effort into the trash bin of forgotten ideas.

For all the excitement of Pearson's laboratory, and for all the mathematical knowledge Tippett developed while there, the problem of the distribution of strength for the weakest fiber remained unsolved. After he returned to the Shirley Institute, Tippett found one of those simple logical truths that lie behind some of the greatest mathematical discoveries. He found a seemingly simple equation that had to connect the distribution of extreme values to the distribution of the sample data.

Being able to write an equation is one thing. Solving it is another matter. He consulted with Pearson who could offer him no help. Over the previous seventy-five years, the engineering profession had developed a large collection of equations and their solutions, which could be found in massive compendiums. Nowhere in these compendiums could Tippett find his equation.

He did what a poor high school algebra student would do. He guessed an answer—and it turned out to solve the equation. Was this the only solution to that equation? Was it even the "correct" answer for his problem? He consulted with R. A. Fisher, who was able to derive Tippett's guess, provide two other solutions, and show that these were the only ones. They are known as "Tippett's three asymptotes of the extreme."

THE DISTRIBUTION OF EXTREMES

What good does it do to know the distribution of extremes? If we know how the distribution of extreme values relates to the distribution of ordinary values, we can keep a record of the heights of yearly floods and predict the most likely height of the hundred-year flood. We can do this because the measurements of the yearly flood values give us enough information to estimate the parameters of Tippett's distributions. Thus, the U.S. Army Corps of Engineers can calculate how high to make the dikes on rivers, and the Environmental Protection Agency can set standards for emissions that will control the extreme values of sudden plumes of gases from industrial chimneys. The cotton industry was able to determine those factors of thread production that influenced the parameters of the distribution of strengths for the weakest fiber.

In 1958, Emil J. Gumbel, then a professor of engineering at Columbia University, published the definitive text on the subject, entitled *Statistics of Extremes*. There have been minor additions to the theory since then, extending the concepts to related situations, but Gumbel's text covers all that a statistician needs to know to deal with this subject. The book includes not only Tippett's original work but later refinements to the theory, many of which were the work of Gumbel himself.

Political Murder

Gumbel has an interesting biography. In the late 1920s and early 1930s he was a beginning-level faculty member at a university in Germany. His early papers indicated that he was a man of great potential, but he had not yet risen to a high level of esteem. As such, his job was far from secure, and his ability to support his wife and children was subject to the whim of government authorities. The Nazis were running rampant in Germany at that time. Although it was officially a political party, the National Socialists were really a party of gangsters. The Brown Shirts were an organization of thugs who enforced party will with threats, beatings, and killings. Anyone who criticized the Nazi party was subject to violent attack, often in the open on city streets, in order to intimidate others. Gumbel had a friend who was attacked and killed in such a public way. There were many witnesses to the murder who could supposedly identify the murderers. However, the court found that there was insufficient evidence to convict, and the Brown Shirts involved went free.

Gumbel was horrified. He attended the trial and saw the way in which the judge dismissed all evidence and made his finding arbitrarily, while the Nazis in the courtroom cheered. Gumbel began examining other cases where murders had been committed in the open and no one was found guilty. He came to the conclusion that the Ministry of Justice had been subverted by the Nazis and that many judges were Nazi sympathizers or were in the pay of the Nazis.

Gumbel collected a number of cases, interviewing the witnesses and documenting the false dismissals of the murderers. In 1922, he published his findings in *Four Years of Political Murder*. He had to arrange for the distribution of his own book, finding that many of the bookstores were afraid to carry it. In the meantime, he continued collecting cases and in 1928 published *Causes of Political Murder*. He tried to create a political group to oppose the Nazis, but most of his fellow academicians were too frightened. Even his Jewish friends were afraid to join him.

When the Nazis came to power in 1933, Gumbel was attending a mathematics meeting in Switzerland. He wanted to rush back to Germany to fight against this new government. His friends dissuaded him, proving to him that he would be arrested and killed before he could barely cross the border. In the early days of the Nazi regime, before the new government was able to control all border crossings, a small number of Jewish professors, like the leading German probabilist Richard von Mises, who foresaw what was to happen, escaped. Gumbel's friends took advantage of this time of confusion to get his family out of Germany, too. They settled briefly in France, but in 1940 the Nazis entered the country.

Gumbel and his family fled south to the unoccupied part of France, where a Nazi-installed rump government ruled, subservient to German demands. Gumbel was among many German democrats whose lives were in danger, since they were on a list of enemies of the state that the Nazis were demanding the French government turn over. Among these other German refugees trapped in Marseilles were Heinrich Mann, the brother of the writer Thomas Mann, and Lion Feuchtwanger. In violation of U.S. State Department regulations, Hiram Bingham IV, the American consul in Marseilles, began issuing visas to these German refugees. He was reprimanded by Washington and finally removed from his post because of his activities. But Bingham was able to save many who faced certain death if the Nazis had had their way. Gumbel and his family came to the United States,[1] where he was offered a position at Columbia University.

[1]On his death in 1966, Gumbel's papers were given to the Leo Baeck Institute of New York, which recently released eight reels of microfilm dealing with his activities against the Nazis, organized under the title *The Emil J. Gumbel Collection, Political Papers of an Anti-Nazi Scholar in Weimar and Exile*.

There are different types of mathematical writing. Some "definitive" texts are cold and sparse, presenting a sequence of theorems and proofs with little or no motivation. In some texts, the proofs are turgid and difficult, plowing with determination from hypotheses to conclusion. There are some definitive texts that are filled with elegant proofs, where the course of the mathematics has been boiled down to seemingly simple steps that move effortlessly to the final conclusions. There are also a very small number of definitive texts in which the authors try to provide the background and ideas behind the problems, in which the history of the subject is described, and in which examples are taken from interesting real-life situations.

These last characteristics describe Gumbel's *Statistics of Extremes*. It is a magnificently lucid presentation of a difficult subject, filled with references to the development of the subject. The first chapter, entitled "Aims and Tools," introduces the topic and develops the mathematics needed to understand the rest of the book. This chapter alone is an excellent introduction to the mathematics of statistical distribution theory. It is designed to be understood by someone with no more background than first-year college calculus. Although I first read the book after I had received my Ph.D. in mathematical statistics, I learned a great deal from that first chapter. In the preface, the author states modestly: "This book is written in the hope, contrary to expectation, that humanity may profit by even a small contribution to the progress of science."

The contribution this book makes could hardly be called "small." It stands as a monument to one of the great teachers of the twentieth century. Emil Gumbel was one of those rare individuals who combine extraordinary courage with an ability to communicate some of the most difficult ideas in a clear and concise fashion.

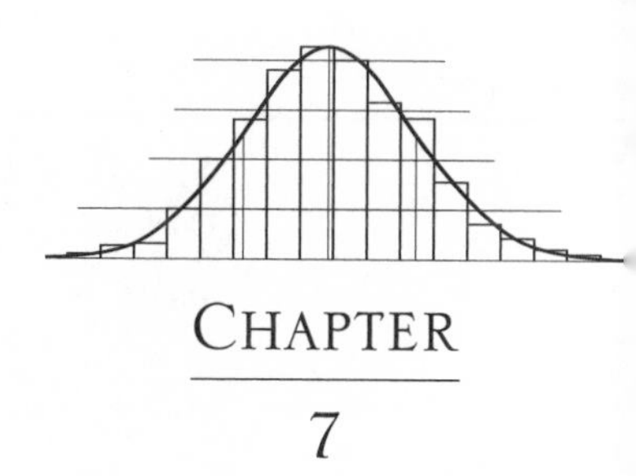

CHAPTER

7

FISHER TRIUMPHANT

The Royal Statistical Society of England publishes articles in its three journals and sponsors meetings throughout each year, at which invited speakers present their latest work. It is difficult to get an article published in one of their journals. It has to be reviewed by at least two referees to see if it is correct; and both an associate editor and the main editor have to agree that it represents a significant increase in natural knowledge. It is an even higher hurdle to be invited to speak at a meeting. This is an honor reserved for only the most prominent workers in the field.

It is the custom of the society to follow each invited talk with a discussion by the members of the audience. Selected members are provided with advance copies of the paper to be presented, so their contributions to the discussion are often detailed and trenchant. *The Journal of the Royal Statistical Society* then publishes both the paper and the comments of the discussants.

The discussion, as it appears in the journal, has a very formal and British tone. The chair of the meeting (or a designate) stands to move a vote of

Ronald Aylmer Fisher, 1890–1962

thanks for the speaker, followed by his comments. A designated senior member of the society then rises to second the vote of thanks, followed by his comments. Then, one by one, some of the most renowned members of the society rise to make their comments. Visitors are often invited from the United States, the Commonwealth, and from other nations, and their comments are included. The speaker responds to these comments. Both the discussants and the speaker are allowed to edit their words before they appear in the journal.

On December 18, 1934, the singular honor of presenting such a paper was given to Professor R. A. Fisher, Sc.D., F.R.S. After his virtual isolation in the 1920s, Fisher was finally having his genius recognized. When we last saw him (in previous chapters), his most senior academic degree was an M.S., and his "university" was a remote agricultural experimental station outside of London. By 1934, he had accumulated an additional degree, doctor of science, and he had been elected a fellow of the prestigious Royal Society (hence the initials F.R.S.). Now, at last, the Royal Statistical Society was granting him a place among the leaders of the field. For this honor, Fisher presented a paper entitled "The Logic of Inductive Inference." The president of the society, Professor M. Greenwood, F.R.S., was in the chair. The printed article runs for sixteen pages and presents a carefully constructed and very clear summary of Fisher's most recent work. The first discussant is Professor A. L. Bowley, who rises to propose a vote of thanks. His comments continue:

> I am glad to have this opportunity of thanking Professor Fisher, not so much for the paper he has read to us, as for his contributions to statistics in general. This is an appropriate occasion to say that I, and all statisticians with whom I associate, appreciate the enormous amount of zeal he has brought to the study of statistics, the power of his mathematical tools, the extent of his influence here, in America and elsewhere, and the stimulus he has given to what he believes to be the correct application of mathematics.

Karl Pearson is not among the discussants. Three years earlier, he had retired from his position at the University of London. Under his leadership, the Galton Biometrical Laboratory had grown into a formal department of biometrics at the university. Upon his retirement, it was split into two departments. Ronald Aylmer Fisher was named chairman of the new Department of Eugenics. Karl Pearson's son, Egon Pearson, was named chairman of the diminished Biometric Department, was in charge of the Galton Biometrical Laboratory, and was editor of *Biometrika*.

Fisher and young Pearson did not have a good personal relationship. This was entirely Fisher's fault. He treated Egon Pearson with obvious hostility. This gentle man suffered from Fisher's dislike of his father and from Fisher's further dislike of Jerzy Neyman, whose collaboration with Egon Pearson will be described in chapter 10. However, young Pearson was most respectful of and had high regard for Fisher's work. In later years, he wrote that he had long since gotten used to Fisher's never having mentioned his name in print. In spite of these tensions and some jurisdictional disputes between the two departments, Fisher and Egon Pearson sent students to each other's lectures and refrained from public disputes.

Karl Pearson, at this time referred to by students as "the old man," was given a single graduate assistant and allowed to maintain an office, but his office was situated in a building a distance from the one that housed the two departments and the Biometrical Laboratory. Churchill Eisenhart, who had come from America to study for a year with Fisher and Egon Pearson, wanted to go over to meet Karl Pearson, but his fellow students and members of the faculty discouraged him. Why, they asked, would anyone want to see Karl Pearson? What did he have to offer to the exciting new ideas and methods that were flowing from the prolific mind of R. A. Fisher? To his regret, Eisenhart never did visit Karl Pearson during his stay in London. Pearson would be dead within the year.

THE FISHERIAN VERSUS THE PEARSONIAN VIEW OF STATISTICS

A philosophical difference separated Karl Pearson's approach to distributions from that of Fisher's. Karl Pearson viewed statistical distributions as describing the actual collections of data he would analyze. According to Fisher, the true distribution is an abstract mathematical formula, and the data collected can be used only to estimate the parameters of the true distribution. Since all such estimates will include error, Fisher proposed tools of analysis that would minimize the degree of such error or that would produce answers that are closer to the truth more often than any other tool. In the 1930s, Fisher appeared to have won the argument. In the 1970s, the Pearsonian view made a comeback. As of this writing, the statistical community is divided over this question, though Pearson would hardly recognize the arguments of his intellectual heirs. Fisher's clear mathematical mind had swept away much of the debris of confusion that kept Pearson from seeing the underlying nature of his views, and later revivals of Pearson's approach have to deal with Fisher's theoretical work. At several places in this book, I plan to look at these philosophical questions, because there are some serious problems with the application of statistical models to reality. This is such a place.

Pearson viewed the distribution of measurements as a real thing. In his approach, there was a large but finite collection of measurements for a given situation. Ideally, the scientist would collect all of these measurements and determine the parameters of their distribution. If all of them could not be collected, then one collected a very large and representative subset of them. The parameters computed from that large, representative subset will be the same as the parameters of the entire collection. Furthermore, the mathematical methods used to compute the values of the parameters for the entire collection can be applied to the representative subset to calculate the parameters without serious error.

To Fisher, the measurements were a random selection from the set of all possible measurements. As a result, any estimate of a parameter based upon that random selection would, itself, be random and have a probability distribution. To keep this idea clear from the idea of an underlying parameter, Fisher called this estimate a "statistic." Modern terminology often calls it an "estimator." Suppose we have two methods of deriving a statistic that estimates a given parameter. For instance, the teacher who wishes to determine how much knowledge a pupil has (the parameter) gives a group of tests (the measurements) and takes the average (the statistic). Would it be "better" to take the median as the statistic, or would it be "better" to take the average of the highest and lowest marks in the group of tests, or would it be "better" still to leave off the highest and lowest marks and use the average of the remaining tests?

Since the statistic is random, it makes no sense to talk about how accurate a single value of it is. This is the same reason it makes no sense to talk about a single measurement and ask how accurate it is. What is needed is a criterion that depends upon the probability distribution of the statistic, just as Pearson proposed that measurements in a set have to be evaluated in terms of their probability distribution and not their individual observed values. Fisher proposed several criteria of a good statistic:

Consistency: The more data you have, the greater the probability that the statistic you calculate is close to the true value of the parameter.

Unbiasedness: If you use a particular statistic many times on different sets of data, the average of those values of the statistic should come close to the true value of the parameter.

Efficiency: Values of the statistic will not be exactly equal to the true value of the parameter, but the bulk of a large number of statistics that estimate a parameter should not be too far from the true value.

These descriptions are a little vague, because I have tried to translate the specific mathematical formulations into plain English. In practice, Fisher's criteria can be evaluated with an appropriate use of mathematics.

Statisticians after Fisher have proposed other criteria. Fisher, himself, proposed some secondary criteria in later work. Cutting away the confusion of all these criteria, the important element is that one considers the statistic to be itself random and that good statistics have good probabilistic properties. Never will we know if the value of a statistic for a particular set of data is correct. We can only say we used a procedure that produces a statistic that meets these criteria.

Of the three fundamental criteria Fisher proposed, the criterion of unbiasedness has caught the public imagination. This is probably because the word *bias* has certain unacceptable overtones. A statistic that is biased seems to be something no one wants. Official guidelines from the U.S. Food and Drug Administration admonish that methods should be used that "avoid bias." A very strange method of analysis (which will be discussed in detail in chapter 27), called "intent to treat," has come to dominate many medical trials because this method guarantees the result will be unbiased, even though it ignores the criterion of efficiency.

In fact, biased statistics are often used with great effectiveness. Following some of Fisher's work, the standard method for determining the concentration of chlorine needed to purify a municipal water supply depends upon a biased (but consistent and efficient) statistic. All of this is some sort of lesson in the sociology of science, how a word created to clearly define a concept has brought its emotional baggage into the science and influenced what people do.

Fisher's Maximum Likelihood Methods

As he worked out the mathematics, Fisher realized that the methods Karl Pearson had been using to compute the parameters of his distributions produced statistics that were not necessarily consistent and were often biased, and that there were much more efficient statistics available. To produce consistent and efficient (but not necessarily unbiased) statistics, Fisher proposed something he called the "maximum likelihood estimator" (MLE).

Fisher then proved that the MLE was always consistent and that (if one allows a few assumptions known as "regularity conditions") it was the most efficient of all statistics. Furthermore, he proved that, if the MLE is biased, the bias can be calculated and subtracted from the MLE, making a modified statistic that is consistent, efficient, and unbiased.[1]

Fisher's likelihood function swept through the mathematical statistics community and rapidly became the major method for estimating parameters. There was only one problem with maximum likelihood estimation. The mathematical problems posed in attempting to solve for the MLEs were formidable. Fisher's papers are filled with line after line of complicated algebra showing the derivation of the MLE for different distributions. His algorithms for analysis of variance and analysis of covariance are magnificent mathematical accomplishments, where he managed to make use of clever substitutions and transformations in multidimensional space to produce formulas that gave the user the MLEs that were needed.

In spite of Fisher's ingenuity, the majority of situations presented intractable mathematics to the potential user of the MLE. The statistical literature of the last half of the twentieth century contains many very clever articles that make use of simplifications of the mathematics to get good approximations of the MLE for certain cases. In my own doctoral thesis (circa 1966), I had to settle for a solution to my problem that was good only if there was a very large amount of data. Assuming that I had a large amount of data enabled me to simplify the likelihood function to a point where I could compute an approximate MLE.

Then came the computer. The computer is not a competitor for the human brain. The computer is just a big, patient number cruncher. It doesn't get bored. It doesn't get sleepy and make a mistake. It will do the same taxing calculation over and over and over and over—for several million "overs"—again. It can find MLEs by methods known as "iterative algorithms."

[1] In the 1950s, C. R. Rao from India and David Blackwell, teaching at Howard University, showed that, if Fisher's regularity conditions do not hold, it is still possible to construct a most efficient statistic from the MLE. The two men, working independently, produced the same theorem and so, in an exception to Stigler's law of misonomy, the Rao-Blackwell theorem does honor its discoverers.

ITERATIVE ALGORITHMS

One of the earliest iterative mathematical methods seems to have emerged during the Renaissance (although David Smith, in his 1923 *History of Mathematics*, claims to have found examples of this method in early Egyptian and Chinese records). The banking companies or counting houses being organized in northern Italy during the first glimmerings of capitalism had a basic problem. Each little city-state or country had its own currency. The counting house had to be able to figure out how to convert the value of, say, a load of lumber, which had been bought for 127 Venetian ducats, to what it should be worth in Athenian drachma, if the exchange rate was 14 drachma to the ducat. Nowadays, we have the power of algebraic notation to get a solution. Remember high school algebra? If x equals the value in drachmas, then . . .

Though mathematicians were beginning to develop algebra at that time, this ease of computation was not available to most people. The bankers used a calculation method called the "rule of false position." Each counting house had its own version of the rule, which was taught to its clerks under a veil of secrecy, because each counting house believed that its version of the rule was "best." Robert Recorde, a sixteenth-century English mathematician, was prominent in popularizing the new algebraic notation. To contrast the power of algebra to the rule of false position, he provides the following version of the rule of false position in *The Grovnd of Artes*, a book he wrote in 1542:

> Gesse at this woorke as happe doth leade.
>
> By chaunce to truthe you may procede.
>
> And firste woorke by the question,
>
> Although no truthe therein be don.
>
> Suche falsehode is so good a grounde,
>
> That truthe by it will soone be founde.
>
> From many bate to many more,

From to fewe take to fewe also.
With to much ioyne to fewe againe,
To to fewe adde to manye plaine.
In crossewaies multiplye contrary kinde,
All truthe by falsehode for to fynde.

What Robert Recorde's sixteenth-century English says is that you first guess the answer and apply it to the problem. There will be a discrepancy between the result of using this guess and the result you want. You take that discrepancy and use it to produce a better guess. You apply this guess and get a new discrepancy, producing another guess. If you are clever in the way you compute the discrepancy, the sequence of guesses will eventually come to the correct answer. For the rule of false position, it takes only one iteration. The second guess is always correct. For Fisher's maximum likelihood, it might take thousands or even millions of iterations before you get a good answer.

What are a mere million iterations to a patient computer? In today's world, it is no more than the blinking of an eye. Not so long ago, computers were less powerful and much slower. In the late 1960s, I had a programmable desk calculator. This was a primitive electronic instrument that would add, subtract, multiply, and divide. But it also had a small memory, where you could put a program that told it to engage in a sequence of arithmetic operations. One of those operations could also change lines of your program. So, it became possible to run an iterative calculation on this programmable calculator. It just took a long time. One afternoon, I programmed the machine, checked the first few steps to make sure I had not made an error in my program, turned off the light in my office, and left for home. Meanwhile, the programmable calculator was adding and subtracting, multiplying and dividing, silently, mumbling away in its electronic innards. Every once in a while it was programmed to print out a result. The printer on the machine was a noisy impact device that made a loud sound like "BRRRAAAK."

The nighttime cleaning crew came into the building and one of the men took his broom and wastepaper collector into my office. There in the darkness, he could hear a humming. He could see the blue light of the calculator's one eye waxing and waning as it added and subtracted over and over again. Suddenly, the machine awoke. "BRRAAK," it said, and then, "BRRAAK, BRRAAK, BRRAAK, BRRRRAAAAK!" He told me later that it was a terrifying experience and asked that I leave some sort of sign up the next time, warning that the computer was at work.

Today's computers work much faster, and ever more complicated likelihoods are being analyzed. Professors Nan Laird and James Ware of Harvard University have invented a remarkably flexible and powerful iterative procedure known as the "EM algorithm." Every new issue of my statistical journals describes how someone has adapted his or her EM algorithm to what had once been considered an unsolvable problem. Other algorithms have been appearing in the literature, going by fanciful names like "simulated annealing" and "kriging." There is the Metropolis algorithm and the Marquardt algorithm, and others named after their discoverers. There are complicated software packages with hundreds and thousands of lines of code that have made such iterative calculations "user friendly."

Fisher's approach to statistical estimation is triumphant. Maximum likelihood rules the world, and Pearson's methods lie in the dust of discarded history. But at that time, in the 1930s—when Fisher was finally recognized for his contributions to the theory of mathematical statistics, when Fisher was in his forties and in the full bloom of his strength—at that very moment, there was a young Polish mathematician named Jerzy Neyman, who was asking questions about some problems that Fisher had swept under the rug.

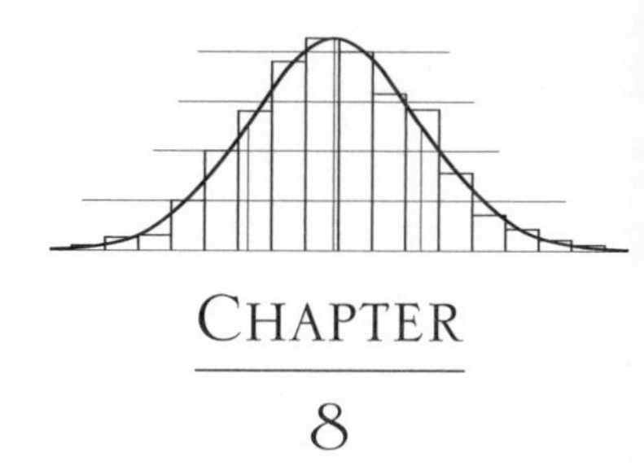

CHAPTER

8

THE DOSE THAT KILLS

Every March, the Biometric Society holds a spring meeting in a southern U.S. city. Those of us who live and work in the North have a chance to come down to Louisville, or Memphis, or Atlanta, or New Orleans, breathe in the new spring air, and watch the flowers and fruit trees bloom a few weeks before it will happen back home. As at other scientific meetings, sessions are held where three to five speakers present papers, and discussants and the audience criticize those papers by questioning the derivations or noting alternative approaches. There are usually two groups of parallel sessions in the morning, a brief break for lunch, followed by two groups of sessions in the afternoon. The last of the sessions is usually over around 5 P.M. The participants return to their rooms in the hotels but gather in groups an hour to an hour and a half later. These small groups then go off to dinner, sampling the restaurants of the city.

You usually meet friends at the sessions and make arrangements for dinner during the day. One day, I had failed to do so. I had gotten into a long, interesting discussion with one of the afternoon speakers. He came from the area and was heading home, so I did not arrange to have dinner with him. When we left off our conversation, the hall was empty, and there was no one else for me to contact. I went back to my room, telephoned my wife, talked to our children, and descended to the hotel lobby. Perhaps I would find a group of people I knew and could join them.

The lobby was almost empty, except for a tall, white-haired man sitting by himself in one of the upholstered chairs. I recognized him as Chester Bliss. I knew him as the inventor of the basic statistical models that were used in determining dose response relationships for drugs and poisons. I had attended a session that morning where he had presented a paper. I went over to him, introduced myself, and complimented him on his talk. He invited me to sit down, and we sat there for a while talking about statistics and mathematics. Yes, it is possible to talk about such things and even make jokes about them. It became obvious that neither of us had made plans for dinner, so we agreed to dine together. He was an entertaining dinner companion, with a wealth of stories from his experience. At meetings in later years, we would sometimes dine together again, and I would often see him when I attended talks sponsored by the Yale University statistics department, where he taught.

Bliss came from a solid middle-American Midwestern home. His father was a doctor. His mother was a homemaker. He had several sisters and brothers. His first interests were in biology, and he studied entomology in college. In the late 1920s, when he graduated, he took a position with the U.S. Department of Agriculture as an entomologist and was soon involved in the development of insecticides. He quickly realized that field experiments with insecticides involved many uncontrolled variables and were difficult to interpret. He took his insects indoors and set up a series of laboratory experiments. Someone introduced him to R. A. Fisher's book, *Statistical Methods for Research Workers*. With this as a start, he found himself reading Fisher's more mathematical papers as he tried to understand what lay behind the methods Fisher showed in the book.

PROBIT ANALYSIS

Soon, following Fisher's lead, Bliss was setting up laboratory experiments in which groups of insects were placed under glass jars and subjected to different combinations and doses of insecticides. As he ran these experiments, he began to notice an interesting phenomenon. No matter how concentrated he made the insecticide, there would always be one or two specimens still alive after exposure. And no matter how weak the pesticide, even if he had used only the liquid carrier, there would be a few insects killed by their exposure.

With this obvious variability, it would be useful to model the effects of insecticides in terms of Pearson's statistical distributions, but how? The reader may recall those terrible moments in high school algebra when the book shifted into word problems. Mr. A and Mr. B were set rowing in still water or against a steady current, or maybe they were mixing water with oil, or bouncing a ball back and forth. Whatever it was, the word problem would propose some numbers and then ask a question, and the poor student had to put those words into a formula and solve for *x*. The reader may recall going back through the pages of the textbook, desperately seeking a similar problem that was worked out as an example and trying to stuff the new numbers into the formulas that were used in that example.

In high school algebra, someone had already worked out the formulas. The teacher knew them or could find them in the teacher's manual for the textbook. Imagine a word problem where nobody knows how to turn it into a formula, where some of the information is redundant and should not be used, where crucial information is often missing, and where there is no similar example worked out earlier in the textbook. This is what happens when one tries to apply statistical models to real-life problems. This was the situation when Chester Bliss tried to adapt the new mathematical ideas of probability distributions to his experiments with insecticides.

Bliss invented a procedure he called "probit analysis." His invention required remarkable leaps of original thought. There was nothing in the works of Fisher, of "Student," or of anyone else that even suggested how he might proceed. He used the word *probit* because his model related the dose to the probability that an insect would die at that dose. The most important parameter his model generated is called the "50 percent lethal dose," usually referred to as the "LD-50." This is the dose of insecticide that has a 50 percent probability of killing. If the insecticide is applied to large numbers of insects, then 50 percent will be killed by the LD-50. Another consequence of Bliss's model is that it is impossible to determine the dose that will kill a specific specimen.

Bliss's probit analysis has been applied successfully to problems in toxicology. In some sense, the insights gained from probit analysis form the foundation of much of the science of toxicology. Probit analysis provides a mathematical foundation for the doctrine first established by the sixteenth-century physician Paracelsus: "Only the dose makes a thing not a poison." Under the Paracelsus doctrine, all things are potential poisons if given in a high enough dose, and all things are nonpoisonous if given in a low enough dose. To this doctrine, Bliss added the uncertainty associated with individual results.

One reason why many foolish users of street drugs die or become very sick on cocaine or heroin or speed is that they see others using the drugs without being killed. They are like Bliss's insects. They look around and see some of their fellow insects still alive. However, knowing that some individuals are still living provides no assurance that a given individual will survive. There is no way of predicting the response of a single individual. Like the individual observations in Pearson's statistical model, these are not the "things" with which science is interested. Only the abstract probability distribution and its parameters (like the LD-50) can be estimated.

Once Bliss had proposed probit analysis,[1] other workers proposed different mathematical distributions. Modern computer programs for computing the LD-50 usually offer the user a selection from several different models that have been proposed as refinements on Bliss's work. Studies using actual data indicate that all these alternatives produce very similar estimates of the LD-50, although they differ in their estimates of doses associated with much lower probabilities, like the LD-10.

It is possible, using probit analysis or any of the alternative models, to estimate a different lethal dose, such as the LD-25 or the LD-80, the doses that will kill 25 percent or 80 percent, respectively. The further you get from the 50 percent point, the more massive the experiment that is needed to get a good estimate. I was once involved in a study to determine the LD-01 of a compound that causes cancer in mice. The study used 65,000 mice, and our analysis of the final results indicated that we still did not have a good estimate of the dose that would produce cancer in 1 percent

---

[1]Stigler's law of misonomy plays a role in probit analysis. Bliss was apparently the first to propose this method of analysis. However, the method required a two-stage iterative calculation and interpolation in a complicated table. In 1953, Frank Wilcoxon at American Cyanamid produced a set of graphs that enabled the user to compute the probit by just laying a ruler across a set of marked lines. This was published in a paper by J. T. Litchfield and Wilcoxon. To prove that this graphical solution produced the correct answer, the authors included an appendix in which they repeated the formulas proposed by Bliss and Fisher. Sometime in the late 1960s, an unknown pharmacologist gave that paper to an unknown programmer, who used the appendix to write a computer program that ran the probit analysis (via Bliss's iterative solution). The documentation of that program used the Litchfield and Wilcoxon paper as its reference. Other probit analysis computer programs soon began appearing at other companies and in academic pharmacology departments, all derived from this original program and all using the Litchfield and Wilcoxon paper as their reference in the documentation. Eventually, the probit analyses run via these programs began to appear in the pharmacological and toxicological literature, and the Litchfield and Wilcoxon paper was used in the references as the "source" of probit analysis. Thus, in the *Science Citation Index*, which tabulates all the references used in most published scientific papers, this paper by Litchfield and Wilcoxon has become one of the most frequently cited in history—not because Litchfield and Wilcoxon did something so great but because Bliss's probit analysis has proved to be so very useful.

of the mice. Calculations based on the data from that study showed that we would need several hundred million mice to get an acceptable estimate of the LD-01.

BLISS IN SOVIET LENINGRAD DURING THE STALIN TERRORS

Chester Bliss's initial work on probit analysis was interrupted in 1933. Franklin D. Roosevelt had been elected president of the United States. In his campaign for the presidency, Roosevelt had made it clear that it was the federal deficit that was causing the depression, and he promised to cut the national deficit and reduce the size of government. That is not what the New Deal eventually did, but that was the campaign promise. When the new president came into office, some of his appointees started firing unnecessary government workers to follow up on this campaign promise. The assistant to the assistant undersecretary of agriculture, in charge of developing new insecticides, looked at what the department had been doing and discovered that someone was foolishly trying to experiment with insecticides inside a laboratory, instead of out in the fields where the insects were. Bliss's laboratory was closed and Bliss was fired. He found himself without a job in the depths of the Great Depression. It did not matter that Chester Bliss had invented probit analysis, and that there were no jobs for an out-of-work entomologist, especially one who worked with insects indoors instead of outdoors where they live.

Bliss contacted R. A. Fisher, who had just taken a new position in London. Fisher offered to put him up and give him some laboratory facilities, but he did not have a job for him and could not pay the American entomologist. Bliss went to England anyway. He lived with Fisher and his family for a few months. Together, he and Fisher refined the methodology of probit analysis. Fisher found some errors in his mathematics and suggested modifications that made the resulting statistics more efficient. Bliss published a new paper, making use of Fisher's suggestions, and Fisher incorporated the necessary tables in a new edition of the book of statistical tables he had written in conjunction with Frank Yates.

After less than a year in England, Fisher found Bliss a job. It was at the Leningrad Plant Institute in the Soviet Union. Imagine tall, thin, Midwestern, middle-American, apolitical Chester Bliss, who was never able to learn a second language, crossing Europe by train with a small suitcase containing his only clothes, and arriving at the station in Leningrad just as the ruthless Soviet dictator, Stalin, was beginning his bloody purges of both major and minor government officials.

Soon after Bliss arrived, the boss of the man who had hired him was called to Moscow—and was never seen again. A month later, the man who had hired Bliss was called to Moscow—and "committed suicide" on the way back. The man in charge of the laboratory next to Bliss's left hastily one day and escaped Russia by sneaking across the Latvian border.

In the meantime, Bliss set to work, treating selected groups of Russian pests with different combinations of insecticides and working out the probits and LD-50s. He rented a room in a house near the institute. His Russian landlady could speak only Russian, and Bliss could speak only English, but he told me that they managed to get along quite well with combinations of gestures and hearty laughs. Bliss met a young woman from America. She had left college to participate in the great Communist experiment in Russia and arrived with the full idealism of youth and the dogmatic blindness of a true Marxist-Leninist. She befriended poor monolingual Chester Bliss and helped him shop and get around the city. She was also a member of the local Communist party. The party knew all about Bliss. They knew when he had been hired, when he arrived in Russia, where he lived, and what he was doing in his lab.

One day, she told him that several members of the party had come to the conclusion that he was an American spy. She had defended him and tried to explain that he was a simple, naive scientist who was only interested in his experiments. However, Moscow had been notified of these suspicions, and they were sending a committee to Leningrad to investigate.

The committee convened at the Leningrad Plant Institute and called Bliss to appear before them to be interrogated. When he walked into the room, he knew who the members of the committee were—having been told by his girlfriend. They had barely gotten through the first few questions, when he said to them, "I see Professor so-and-so is among you [Bliss could not remember his name when he told me the story]. I have been reading his papers. Tell me, this method of agricultural experimentation he proposes, is it the gospel according to Saint Marx and Saint Lenin?" The interpreter was hesitant to translate this question, and when he did there was some stirring among the committee members. They asked him to elaborate.

"Professor so-and-so's method," Bliss asked, "is this the official party line? Is this the way in which the party requires that agricultural experimentation be done?"

The eventual response was, yes, this was the correct way to do things.

"Well, in that case," Bliss said, "I am in violation of your religion." Bliss went on to explain that the methods of agricultural research proposed by this man required that vast tracts of land all be given the same treatment. Bliss said that he considered such experiments useless and pointed out that he had been advocating the use of small neighboring plots with different treatments assigned to rows within those plots.

The interrogation did not proceed much further. That evening, Bliss's friend told him the committee had decided he was not a spy. He was too open and obvious. He was probably what she said he was, a simpleminded scientist going about his experiments.

Bliss continued working at the Leningrad Plant Institute for the next few months. He had no boss anymore. He just did what he thought best. He had to join the Communist labor union of laboratory workers. Everyone who had a job in Russia was forced to belong to a government-run labor union. Other than that, they left him alone. In the 1950s, the U.S. State Department would deny him his passport because he had once belonged to a Communist organization.

One afternoon, his girlfriend burst into his lab. "You have to leave immediately," she told him. He protested that his experiments had not been finished, that he had not yet written up his notes. She pushed him away from his notebooks and started getting him his coat. He had to leave without delay. He had to abandon everything, she told him. She watched him as he packed his small suitcase and said goodbye to his landlady. His friend saw him to the train station, and she insisted that he telephone her when he was safely in Riga.

In the early 1960s, the cold hand of repression was lifted slightly in the Soviet Union. Soviet scientists rejoined the international scientific community, and the International Statistical Institute (of which Chester Bliss was a fellow) arranged a meeting in Leningrad. Between sessions, Bliss set out to look up his old friends from the 1930s. They were all dead, having been murdered during Stalin's purges or killed during World War II. Only his landlady remained alive. The two of them greeted each other, bobbing and nodding, he muttering his good wishes in English and she responding in Russian.

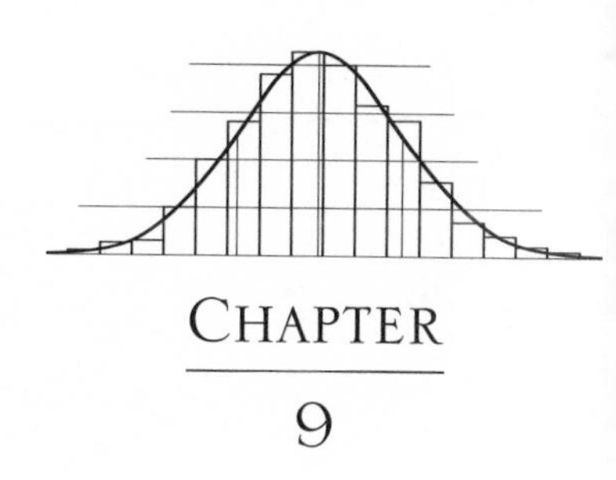

CHAPTER

9

THE BELL-SHAPED CURVE

The reader of the first eight chapters of this book has probably come to believe that the statistical revolution occurred solely in Great Britain. To some extent this is true, in that the first attempts to apply statistical models to biological and agricultural studies were in Great Britain and also in Denmark. Under R. A. Fisher's influence, statistical methods soon spread to the United States, India, Australia, and Canada. While the immediate applications of statistical models were being pushed in the English-speaking world, continental Europe had a long mathematical tradition, and European mathematicians were working on the theoretical problems related to statistical modeling.

Foremost among these was the central limit theorem. Until the early 1930s, this was an unproven theorem, a conjecture that many believed was true but which no one had been able to prove. Fisher's theoretical work on the value of the likelihood function assumed this theorem to be true. Pierre Simon Laplace, in the early nineteenth century, had justified his method of least squares with that assumption. The new science of psychology developed techniques of measuring intelligence and scales of mental illness that rested on the central limit theorem.

WHAT IS THE CENTRAL LIMIT THEOREM?

The averages of large collections of numbers have a statistical distribution. The central limit theorem states that this distribution can be approximated by the normal probability distribution regardless of where the initial data came from. The normal probability distribution is the same as Laplace's error function. It is sometimes called the "Gaussian distribution." It has been described loosely in popular works as the "bell-shaped curve." In the late eighteenth century, Abraham de Moivre proved that the central limit theorem holds for simple collections of numbers from games of chance. For the next 150 years, no further progress was made on proving this conjecture.

The conjecture was widely assumed true because it justified using the normal distribution to describe most data. Once one assumes that the normal distribution is involved, the mathematics become much more tractable. The normal distribution has some very nice properties. If two random variables have a normal distribution, their sum has a normal distribution. In general, all kinds of sums and differences of normal variables have normal distributions. So, many statistics derived from normal variates are themselves normally distributed.

The normal distribution has only two of Karl Pearson's four parameters, the mean and the standard deviation. The symmetry and kurtosis are equal to zero. Once these two numbers are known, everything else is known. Fisher showed that estimates of the mean and the standard deviation taken from a set of data are what he called sufficient. They contain all the information in the data. There is no need to keep any of the original measurements, since these two numbers contain all that can be discovered from those measurements. If there are enough measurements to allow for reasonably precise estimates of the mean and the standard deviation, no more measurements are needed, and the effort to collect them is a waste of time. For instance, if you are willing to settle for knowing the two parameters of a normal distribution within two significant figures, you need collect only about fifty measurements.

The mathematical tractability of the normal distribution means that the scientist can pose a complicated model of relationships. As long as the underlying distribution is normal, Fisher's likelihood function often has a form that can be manipulated with simple algebra. Even for models that are so complicated they need iterative solutions, Nan Laird and James Ware's EM algorithm becomes especially easy to use if the distributions are normal. In modeling problems, statisticians often act as if all the data are normally distributed because the mathematics are so tractable. To do so, however, they have to invoke the central limit theorem.

But, was the central limit theorem true? To be more exact, under what conditions was it true?

In the 1920s and 1930s, a group of mathematicians in Scandinavia, Germany, France, and the Soviet Union was pursuing these questions with a set of new mathematical tools that had been discovered in the early years of the twentieth century. This, in the face of an impending disaster for all of civilization—the rise of evil totalitarian states.

A mathematician does not need a laboratory with expensive equipment. In the 1920s and 1930s, the typical equipment of a mathematician was a blackboard and chalk. It is better to do mathematics on a chalkboard than on a piece of paper because chalk is easier to erase, and mathematical research is always filled with mistakes. Very few mathematicians can work alone. If you are a mathematician, you need to talk about what you are doing. You need to expose your new ideas to the criticism of others. It is so easy to make mistakes or to include hidden assumptions that you do not see, but that are obvious to others. There is an international community of mathematicians who exchange letters, go to meetings, and examine each other's papers, constantly criticizing, questioning, exploring ramifications. In the early 1930s, William Feller and Richard von Mises in Germany, Paul Lévy in France, Andrei Kolmogorov in Russia, Jarl Waldemar Lindeberg and Harald Cramér in Scandinavia, Abraham Wald and Herman Hartley in Austria, Guido Castelnuovo in Italy, and many others were all in communication, many of them examining the central limit conjecture with these new tools.

This free and easy interaction, however, was soon to cease. The dark shadows of Stalin's terror, Nazi racial theories, and Mussolini's dreams of empire were about to destroy it. Stalin was perfecting a combination of rigged show trials and arrests in the middle of the night, killing and intimidating anyone who fell under his paranoid suspicion. Hitler and his criminal henchmen were dragging Jewish professors (primarily) out of the universities and into brutal work camps. Mussolini was locking people into preordained castes that he called the "corporate state."

VIVA LA MUERTE!

An extreme example of this rampant anti-intellectualism occurred during the Spanish Civil War, where the twin evils of fascism and Stalinism were fighting a vicious proxy war with the lives of brave young Spaniards. The Falangists (as the Spanish fascists were known) had conquered the ancient University of Salamanca. The rector of the university was the world-renowned Spanish philosopher Miguel de Unamuno, then in his early seventies. The Falangist general Millan Astray, who had lost a leg, an arm, and an eye in a previous war, was the propaganda chief of the newly conquering forces. His motto was *Viva la muerte!* (Long live death!). Like Shakespeare's King Richard III, Millan Astray's crippled body was a metaphor for his twisted and evil mind. The Falangists decreed a great celebration in the ceremonial hall of the University of Salamanca. The dais held the newly appointed governor of the province, Señora Francisco Franco, Millan Astray, the bishop of Salamanca, and an elderly Miguel de Unamuno, dragged forth as a trophy of their conquests.

Viva la muerte! Millan Astray shouted, and the crowded hall echoed his cry. *España*! someone shouted, and the hall responded, *España! Viva la muerte!* The Falangists in their blue uniforms stood up in unison and gave a fascist salute to the portrait of Franco above the dais. In the midst of these cries, Unamuno stood up and slowly moved to the podium. He began quietly:

> All of you are hanging on my words. You all know me and are aware that I am unable to remain silent. At times to be silent is to lie. For silence can be interpreted as acquiescence. I want to comment on the speech—to give it a name—of General Millan Astray. . . . Just now I heard a necrophilous and senseless cry: "Long live death." And I, who have spent my life shaping paradoxes . . . I must tell you, as an expert authority, that this outlandish paradox is repellent to me. General Millan Astray is a cripple. . . . He is a war cripple. . . . Unfortunately, there are too many cripples in Spain just now. And soon there will be even more of them if God does not come to our aid. . . .

Millan Astray pushed Unamuno aside and cried out, *Abajo la inteligencia! Viva la muerte!* Echoing his cries, the Falangists pushed forward to seize Unamuno, but the old rector continued:

> This is the temple of the intellect. And I am its high priest. It is you who profane its sacred precincts. You will win, because you have more than enough brute force. But you will not convince. For to convince you need to persuade. And in order to persuade you would need what you lack: reason and right. . . .

Unamuno was forced into house arrest and was declared "dead of natural causes" within the month.

The Stalin terror began to cut off communication between the Russian mathematicians and the rest of Europe. Hitler's racial policies decimated German universities, since many of the great European mathematicians were Jewish or were married to Jews, and most of those who were not Jewish were opposed to the Nazi plans. William Feller came to Princeton University, Abraham Wald to Columbia University. Herman Hartley and Richard von Mises went to London. Emil J. Gumbel escaped to France. Emmy Noether was given a temporary position on the faculty of Bryn Mawr College in Pennsylvania.

But not everyone escaped. American immigration gates were closed to anyone who could not show that he or she had employment waiting in the United States. Latin American nations opened and closed their doors on the whims of petty bureaucrats. When the Nazi forces conquered Warsaw, they hunted down all the members of the faculty of the University of Warsaw they could find, brutally murdered them, and buried them in a mass grave. In the Nazi racial world, Poles and other Slavs were to be the uneducated slaves of their Aryan masters. Many of the young and promising students of the ancient universities of Europe perished. In the Soviet Union, the major mathematicians sought the refuge of pure mathematics, with no attempt at applications, since it was in the applications that scientists fell under Stalin's cold suspicions.

Before these shadows had all become reality, however, European mathematicians solved the problem of the central limit theorem. Jarl Waldemar Lindeberg from Finland and Paul Lévy from France independently discovered a set of overlapping conditions that were required for the conjecture to be true. It turned out that there were at least three different approaches to the problem and that there was not a single theorem but a group of central limit theorems, each derived from a slightly different set of conditions. By 1934, the central limit theorem(s) was no longer a conjecture. All one had to do was prove that the Lindeberg-Lévy conditions held. Then the central limit theorem holds, and the scientist is free to assume the normal distribution as an appropriate model.

FROM LINDEBERG-LÉVY TO U-STATISTICS

However, it is difficult to prove that the Lindeberg-Lévy conditions hold for a particular situation. There is a certain comfort in knowing about the Lindeberg-Lévy conditions, because they describe conditions that seem reasonable and are probably true in most situations. But proving them is a different matter. This is why Wassily Hoeffding, toiling away at the University of North Carolina after the war, is so important to this story. In 1948, Hoeffding published a paper, "A Class of Statistics with Asymptotically Normal Distribution," in the *Annals of Mathematical Statistics*.

Recall that R. A. Fisher defined a *statistic* as a number that is derived from the observed measurements and that estimates a parameter of the distribution. Fisher established some criteria that a statistic should have in order to be useful, showing in the process that many of Karl Pearson's methods led to statistics that did not meet these criteria. There are different ways of computing statistics, many of which meet Fisher's criteria. Once the statistic is computed, we have to know its distribution to use it. If it has a normal distribution, it is much easier to use. Hoeffding showed that a statistic that is part of a class he called "U-statistics" meets the Lindeberg-Lévy conditions. Since this is so, one need only show that a new statistic meets Hoeffding's definition and not have to work out the difficult mathematics to show Lindeberg-Lévy holds true. All Hoeffding did was to replace one set of mathematical requirements with another. However, Hoeffding's conditions are, in fact, very easy to check. Since the publication of Hoeffding's paper, almost all articles that show that a new statistic has a normal distribution do so by showing that the new statistic is a U-statistic.

HOEFFDING IN BERLIN

Wassily Hoeffding had an ambiguous situation during World War II. Born in Finland in 1914, to a Danish father and Finnish mother, at a time when Finland was part of the Russian Empire, Hoeffding moved with his family to Denmark and then Berlin after World War I. He thus had dual citizenship in two Scandinavian countries. He finished high school in 1933 and began studying mathematics in Berlin, just as the Nazis came to power in Germany. Anticipating what would happen, Richard von Mises, the head of the mathematics department at his university, left Germany early. Many of Hoeffding's other professors fled soon after or were removed from their positions. In the confusion, young Hoeffding took courses with lower-level instructors, many of whom did not last to complete the courses they were teaching, as the Nazis continued to "cleanse" the faculties of Jews and Jewish sympathizers.

Along with the other math students, he was forced to attend a lecture given by Ludwig Bieberbach, a hitherto minor member of the faculty whose enthusiastic espousal of the Nazi party made him the new department chairman. Bieberbach's lecture dealt with the difference between "Aryan" and "non-Aryan" mathematics. He found that the decadent "non-Aryan" (read Jewish) mathematicians depended upon complicated algebraic notation, whereas "Aryan" mathematicians worked in the nobler and more pure realm of geometric intuition. At the end of the lecture, he called for questions, and a student in the back row asked him why it was that Richard Courant (one of the great Jewish mathematicians in early-twentieth-century Germany) used geometric insights to develop his theories of real analysis. Bieberbach never gave another public lecture on the subject. However, he founded the journal *Deutsche Mathematik*, which soon became the primary mathematics journal in the eyes of the authorities.

Hoeffding finished his studies at the university in 1940, at an age when other young men were being conscripted into the army.

However, his ambiguous citizenship and the fact that Finland was an ally of Germany meant that he was exempted. He took a job as a research assistant with an interuniversity institute for actuarial science. He also worked part-time in the offices of one of the older German mathematics journals, a journal that, unlike Bieberbach's, had difficulty getting paper and was published infrequently. Hoeffding did not even seek a teaching job, since he would need to apply for formal German citizenship to be eligible.

In 1944, non-German citizens "of German or related blood" were declared subject to military service. However, on his physical examination, Hoeffding was found to have diabetes and was excused from the army. He was now eligible for labor service. Harald Geppert, the editor of the journal where he worked part-time, suggested that he do some sort of mathematical work with military applications. He made the suggestion while another editor, Hermann Schmid, was in the room. Hoeffding hesitated and then, trusting of Geppert's discretion, told Geppert that any kind of war work would be contrary to his conscience. Schmid was a member of a noble Prussian family, and Hoeffding hoped that his sense of honor would keep this conversation private.

Wassily Hoeffding sweated out the next few days, but nothing happened to him, and he was allowed to continue his work. As the Russian army approached, Geppert fed poison to his young son for breakfast one morning and then he and his wife took poison. In February 1945, Hoeffding fled with his mother to a small town in Hannover, and were there when it became part of the British zone of occupation. His father remained behind in Berlin, where he was captured by the Russian secret police who considered him a spy, because he had once worked for the American commercial attaché in Denmark. The family remained ignorant of his fate for several years, until his father managed to escape from prison and make his way to the West. In the meantime, young Hoeffding came to New York in the fall of 1946 to continue his studies and was later invited to join the faculty at the University of North Carolina.

Operations Research

One consequence of the Nazi's anti-intellectualism and anti-Semitism was that the World War II Allies reaped a harvest of brilliant scientists and mathematicians to help in their war effort. The British biologist Peter Blackett proposed to the Admiralty that the armed forces could use scientists to solve their strategic and tactical problems. Scientists, regardless of their field of expertise, are trained to apply logic and mathematical models to problems. He proposed that teams of scientists be put to work on war-related problems. Thus was born the discipline of operational research (called operations research in the United States). Teams of scientists from different fields combined to determine the best use of long-range bombers against submarines, to provide firing tables for antiaircraft guns, to determine the best placement of ammunition depots behind front lines, and even to solve questions involving food supplies for troops.

Operations research moved from the battlefield to the business world at the end of the war. The scientists who enlisted in the war showed how mathematical models and scientific thinking could be used to solve tactical problems in warfare. The same approach and many of the same methods could be used in organizing work on a factory floor, finding optimum relationships between warehouses and salesrooms, and in solving many other business problems that involved balancing limited resources or improving production and output. Since that time, operations research departments have been established in most large corporations. Much of the work done by these departments involves statistical models. While at Pfizer, Inc., I worked on several projects aimed at improving the way drug research was managed and new products were brought forward for testing. An important tool in all this work is the ability to call forth the normal distribution when conditions warrant it.

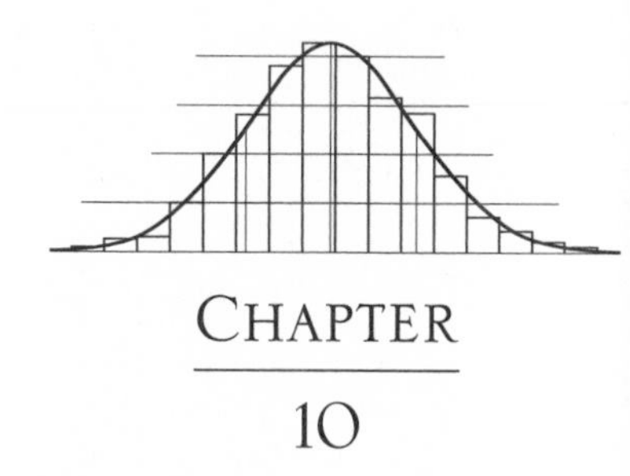

CHAPTER

10

TESTING THE GOODNESS OF FIT

During the 1980s a new type of mathematical model emerged and caught the public imagination, primarily because of its name: "chaos theory."[1] The name suggests some form of statistical modeling with a particularly wild type of randomness. The people who coined the name purposely stayed away from using the word *random*. Chaos theory is actually an attempt to undo the statistical revolution by reviving determinism at a more sophisticated level.

Recall that, before the statistical revolution, the "things" with which science dealt were either the measurements made or the physical events that generated those measurements. With the statistical revolution, the things of science became the parameters that governed the distribution of the measurements. In the earlier deterministic approach, there was always the belief that more refined

[1]The descriptions of chaos theory used here are taken from Brian Davies, *Exploring Chaos: Theory and Experiment* (Reading, MA: Perseus Books, 1999).

measurements would lead to a better definition of the physical reality being examined. In the statistical approach, the parameters of the distribution are sometimes not required to have a physical reality and can only be estimated with error, regardless of how precise the measuring system. For example, in the deterministic approach, there is a fixed number, the gravitational constant, that describes how things fall to the Earth. In the statistical approach, our measurements of the gravitational constant will always differ from one another, and the scatter of their distribution is what we wish to establish in order to "understand" falling bodies.

In 1963, the chaos theorist Edward Lorenz presented an often-referenced lecture entitled "Does the Flap of a Butterfly's Wings in Brazil Set Off a Tornado in Texas?" Lorenz's main point was that chaotic mathematical functions are very sensitive to initial conditions. Slight differences in initial conditions can lead to dramatically different results after many iterations. Lorenz believed that this sensitivity to slight differences in the beginning made it impossible to determine an answer to his question. Underlying Lorenz's lecture was the assumption of determinism, that each initial condition can theoretically be traced as a cause of a final effect. This idea, called the "Butterfly Effect," has been taken by the popularizers of chaos theory as a deep and wise truth.

However, there is no scientific proof that such a cause and effect exists. There are no well-established mathematical models of reality that suggest such an effect. It is a statement of faith. It has as much scientific validity as statements about demons or God. The statistical model that defines the quest of science in terms of parameters of distributions is also based on a statement of faith about the nature of reality. My experience in scientific research has led me to believe that the statistical statement of faith is more likely to be true than the deterministic one.

CHAOS THEORY AND GOODNESS OF FIT

Chaos theory results from the observation that numbers generated by a fixed deterministic formula can appear to have a random pattern. This was seen when a group of mathematicians took some relatively simple iterative formulas and plotted the output. In chapter 9, I described an iterative formula as one that produces a number, then uses that number in its equations to produce another number. The second number is used to produce a third number, and so on. In the early years of the twentieth century, the French mathematician Henri Poincaré tried to understand complicated families of differential equations by plotting successive pairs of such numbers on a graph. Poincaré found some interesting patterns in those plots, but he did not see how to exploit those patterns and dropped the idea. Chaos theory starts with these Poincaré plots. What happens as you build a Poincaré plot is that the points on the graph paper appear, at first, as if they have no structure to them. They appear in various places in a seemingly haphazard fashion. As the number of points in the plot increases, however, patterns begin to emerge. They are sometimes groups of parallel straight lines. They might also be a set of intersecting lines, or circles, or circles with straight lines across them.

The proponents of chaos theory suggest that what in real life appear to be purely random measurements are, in fact, generated by some deterministic set of equations, and that these equations can be deduced from the patterns that appear in a Poincaré plot. For instance, some proponents of chaos theory have taken the times between human heartbeats and put them into Poincaré plots. They claim to see patterns in these plots, and they have found deterministic generating equations that appear to produce the same type of pattern.

As of this writing, there is one major weakness to chaos theory applied in this fashion. There is no measure of how good the fit is between the plot based on data and the plot generated by a specific set of equations. The proof that the proposed generator is correct is based on asking the reader to look at two similar graphs. This eyeball test has proved to be a fallible one in statistical analysis. Those things that seem to the eye to be similar or very close to the same are often drastically different when examined carefully with statistical tools developed for this purpose.

PEARSON'S GOODNESS OF FIT TEST

This problem was one that Karl Pearson recognized early in his career. One of Pearson's great achievements was the creation of the first "goodness of fit test." By comparing the observed to the predicted values, Pearson was able to produce a statistic that tested the goodness of the fit. He called his test statistic a "chi square goodness of fit test." He used the Greek letter chi (χ), since the distribution of this test statistic belonged to a group of his skew distributions that he had designated the chi family. Actually, the test statistic behaved like the square of a chi, thus the name "chi squared." Since this is a statistic in Fisher's sense, it has a probability distribution. Pearson proved that the chi square goodness of fit test has a distribution that is the same, regardless of the type of data used. That is, he could tabulate the probability distribution of this statistic and use that same set of tables for every test. The chi square goodness of fit test has a single parameter, which Fisher was to call the "degrees of freedom." In the 1922 paper in which he first criticized Pearson's work, Fisher showed that, for the case of comparing two proportions, Pearson had gotten the value of that parameter wrong.

But just because he made a mistake in one small aspect of his theory is no reason to denigrate Pearson's great achievement. Pearson's goodness of fit test was the forerunner of a major component of modern statistical analysis. This component is called "hypothesis testing," or "significance testing." It allows the analyst to propose two or more competing mathematical models for reality and use the data to reject one of them. Hypothesis testing is so widely used that many scientists think of it as the only statistical procedure available to them. The use of hypothesis testing, as we shall see in later chapters, involves some serious philosophical problems.

Testing Whether the Lady Can Taste a Difference in the Tea

Suppose we wish to test whether the lady can detect the difference between a cup of tea into which the milk has been poured into the tea versus a cup of tea wherein the tea has been poured into the milk. We present her with two cups, telling her that one of the cups is tea into milk and the other is milk into tea. She tastes and identifies the cups correctly. She could have done this by guessing; she had a 50:50 chance of guessing correctly. We present her with another pair of the same type. Again, she identifies them correctly. If she were just guessing, the chance of this happening twice in a row is $1/4$. We present her with a third pair of cups, and again she identifies them correctly. The chance that this has happened as a result of pure guesswork is $1/8$. We present her with more pairs, and she keeps identifying the cups correctly. At some point, we have to be convinced that she can tell the difference. Suppose she was wrong with one pair. Suppose further that this was the twenty-fourth pair and she was correct on all the others. Can we still conclude that she is able to detect a difference? Suppose she was wrong in four out of the twenty-four? Five of the twenty-four?

Hypothesis, or significance, testing is a formal statistical procedure that calculates the probability of what we have observed, assuming that the hypothesis to be tested is true. When the observed probability is very low, we conclude that the hypothesis is not true. One important point is that hypothesis testing provides a tool for rejecting a hypothesis. In the case above, this is the hypothesis that the lady is only guessing. It does not allow us to accept a hypothesis, even if the probability associated with that hypothesis is very high.

Somewhere early in the development of this general idea, the word *significant* came to be used to indicate that the probability was low enough for rejection. Data became significant if they could be used to reject a proposed distribution. The word was used in its late-nineteenth-century English meaning, which is simply that the computation signified or showed something. As the English language entered the twentieth century, the word *significant* began to take on other meanings, until it developed its current meaning, implying something very important. Statistical analysis still uses the word significant to indicate a very low probability computed under the hypothesis being tested. In that context, the word has an exact mathematical meaning. Unfortunately, those who use statistical analysis often treat a significant test statistic as implying something much closer to the modern meaning of the word.

FISHER'S USE OF P-VALUES

R. A. Fisher developed most of the significance testing methods now in general use. He referred to the probability that allows one to declare significance as the "p-value." He had no doubts about its meaning or usefulness. Much of *Statistical Methods for Research Workers* is devoted to showing how to calculate p-values. As I noted earlier, this was a book designed for nonmathematicians who want to use statistical methods. In it, Fisher does not describe how these tests were derived, and he never indicates exactly what p-value one might call significant. Instead, he displays examples of calculations and notes whether the result is significant or not. In one example, he shows that the p-value is less than .01 and states: "Only one value in a hundred will exceed [the calculated test statistic] by chance, so that the difference between the results is clearly significant."

The closest he came to defining a specific p-value that would be significant in all circumstances occurred in an article printed in the *Proceedings of the Society for Psychical Research* in 1929. Psychical research refers to attempts to show, via scientific methods, the existence of clairvoyance. Psychical researchers make extensive use of statistical significance tests to show that their results are improbable in terms of the hypothesis that the results are due to purely random guesses by the subjects. In this article, Fisher condemns some writers for failing to use significance tests properly. He then states:

> In the investigation of living beings by biological methods, statistical tests of significance are essential. Their function is to prevent us being deceived by accidental occurrences, due not to the causes we wish to study, or are trying to detect, but to a combination of many other circumstances which we cannot control. An observation is judged significant, if it would rarely have been produced, in the absence of a real cause of the kind we are seeking. It is a common practice to judge a result significant, if it is of such a magnitude that it would have been produced by chance not more frequently than once in twenty trials. This is an arbitrary, but convenient, level of significance for the practical investigator, but it does not mean that he allows himself to be deceived once in every twenty experiments. The test of significance only tells him what to ignore, namely all experiments in which significant results are not obtained. He should only claim that a phenomenon is experimentally demonstrable when he knows how to design an experiment so that it will rarely fail to give a significant result. Consequently, isolated significant results which he does not know how to reproduce are left in suspense pending further investigation.

Note the expression "knows how to design an experiment . . . that . . . will rarely fail to give a significant result." This lies at the heart of Fisher's use of significance tests. To Fisher, the significance test makes sense only in the context of a sequence of experiments, all aimed at elucidating the effects of specific treatments. Reading through Fisher's applied papers, one is led to believe that he used significance tests to come to one of three possible conclusions. If the p-value is very small (usually less than .01), he declares that an effect has been shown. If the p-value is large (usually greater than .20), he declares that, if there is an effect, it is so small that no experiment of this size will be able to detect it. If the p-value lies in between, he discusses how the next experiment should be designed to get a better idea of the effect. Except for the above statement, Fisher was never explicit about how the scientist should interpret a p-value. What seemed to be intuitively clear to Fisher may not be clear to the reader.

We will come back to reexamine Fisher's attitude toward significance testing in chapter 18. It lies at the heart of one of Fisher's great blunders, his insistence that smoking had not been shown to be harmful to health. But let us leave Fisher's trenchant analysis of the evidence involving smoking and health for later and turn to 35-year-old Jerzy Neyman in the year 1928.

JERZY NEYMAN'S MATHEMATICAL EDUCATION

Jerzy Neyman was a promising mathematics student when World War I erupted across his homeland in Eastern Europe. He was driven into Russia, where he studied at the University of Kharkov, a provincial outpost of mathematical activity. Lacking teachers who were up to date in their knowledge, and forced to miss semesters of schooling because of the war, he took the elementary mathematics he was taught at Kharkov and built upon it by seeking out articles in the mathematics journals available to him. Neyman thus received a formal mathematics education similar to that taught to students of the nineteenth century, and then educated himself into twentieth-century mathematics.

The journal articles available to Neyman were limited to what he could find in the libraries of the University of Kharkov and later at provincial Polish schools. By chance, he came across a series of articles by Henri Lebesgue of France. Lebesgue (1875–1941) had created many of the fundamental ideas of modern mathematical analysis in the early years of the twentieth century, but his papers are difficult to read. Lebesgue integration, the Lebesgue convergence theorem, and other creations of this great mathematician have all been simplified and organized in more understandable forms by later mathematicians. Nowadays, no one reads Lebesgue in the original. Students all learn about his ideas through these later versions.

No one, that is, except Jerzy Neyman, who had only Lebesgue's original articles, who struggled through them, and who emerged seeing the brilliant light of these great new (to him) creations. For years afterward, Neyman idolized Lebesgue, and, in the late 1930s, finally got to meet him at a mathematics conference in France. According to Neyman, Henri Lebesgue turned out to be a gruff, impolite man, who responded to Neyman's enthusiasm with a few mutterings and turned and walked away in the midst of Neyman's talking to him.

Neyman was deeply hurt by this rebuff, and perhaps with this as an object lesson, always went out of his way to be polite and kind to young students, to listen carefully to what they said, and to engage them in their enthusiasms. That was Jerzy Neyman, the man. All who knew him remember him for his kindness and caring manners. He was gracious and thoughtful and dealt with people with genuine pleasure. When I met him, he was in his early eighties, a small, dignified, well-groomed man with a neat white moustache. His blue eyes sparkled as he listened to others and engaged in intensive conversation, giving the same personal attention to everyone, no matter who they were.

In the early years of his career, Jerzy Neyman managed to find a position as a junior member of the faculty of the University of Warsaw. At that time, the newly independent nation of Poland had little money to support academic research, and positions for mathematicians were scarce. In 1928, he spent a summer at the biometrical laboratory in London and there came to know Egon S. Pearson and his wife, Eileen, and their two daughters. Egon Pearson was Karl Pearson's son, and a more striking contrast in personalities is hard to find. Where Karl Pearson was driving and dominating, Egon Pearson was shy and self-effacing. Karl Pearson rushed through new ideas, often publishing an article with the mathematics vaguely sketched in or even with some errors. Egon Pearson was extremely careful, worrying over the details of each calculation.

The friendship between Egon Pearson and Jerzy Neyman is preserved in their exchange of letters between 1928 and 1933. These letters provide a wonderful insight into the sociology of science, showing how two original minds grapple with a problem, each one proposing ideas or criticizing the ideas of the other. Pearson's self-effacing comes to the forefront as he hesitantly suggests that perhaps something Neyman had proposed might not work out. Neyman's great originality comes out as he cuts through complicated problems to find the essential nature of each difficulty. For someone who wants to understand why mathematical research is so often a cooperative venture, I recommend the Neyman-Pearson letters.

What was the problem that Egon Pearson first proposed to Neyman? Recall Karl Pearson's chi square goodness of fit test. He developed it to test whether observed data fit a theoretical distribution. There really is no such thing as *the* chi square goodness of fit test. The analyst has available an infinite number of ways to apply the test to a given set of data. There appeared to be no criterion on how "best" to pick among those many choices. Every time the test is applied, the analyst must make arbitrary choices. Egon Pearson posed the following question to Jerzy Neyman:

> If I have applied a chi square goodness of fit test to a set of data versus the normal distribution, and if I have failed to get a significant p-value, how do I know that the data really fit a normal distribution? That is, how do I know that another version of the chi square test or another goodness of fit test as yet undiscovered might not have produced a significant p-value and allowed me to reject the normal distribution as fitting the data?

NEYMAN'S STYLE OF MATHEMATICS

Neyman took this question back to Warsaw with him, and the exchange of letters began. Both Neyman and young Pearson were impressed with Fisher's concept of estimation based on the likelihood function. They began their investigation by looking at the likelihood associated with a goodness of fit test. The first of their joint papers describes the results of those investigations. It is the most difficult of the three classic papers they produced, which were to revolutionize the whole idea of significance testing. As they continued looking at the question, Neyman's great clarity of vision kept distilling the problem down to its essential elements, and their work became clearer and easier to understand.

Although the reader may not believe it, literary style plays an important role in mathematical research. Some mathematical writers seem unable to produce articles that are easy to understand. Others seem to get a perverse pleasure out of generating many lines of symbolic notation so filled with detail that the general idea is lost in the picayune. But some authors have the ability to display complicated ideas with such force and simplicity that the development appears to be obvious in their exposition. Only upon reviewing what has been learned does the reader realize the great power of the results. Such an author was Jerzy Neyman. It is a pleasure to read his papers. The ideas evolve naturally, the notation is deceptively simple, and the conclusions appear to be so natural that you find it hard to see why no one produced these results long before.

Pfizer Central Research, where I worked for twenty-seven years, sponsors a yearly colloquium at the University of Connecticut. The statistics department of the university invites a major figure in biostatistical research to come for a day, meet with students, and then present a talk in the late afternoon. Since I was involved in setting up the grant for this series, I had the honor of meeting some of the great men of statistics through them. Jerzy Neyman was one such invitee. He asked that his talk have a particular form. He wanted to present a paper and then have a panel of discussants who would criticize his paper. Since this was the renowned Jerzy Neyman, the organizers of the symposium contacted well-known senior statisticians in the New England area to constitute the panel. At the last minute, one of the panelists was unable to come, and I was asked to substitute for him.

Neyman had sent us a copy of the paper he planned to present. It was an exciting development, wherein he applied work he had done in 1939 to a problem in astronomy. I knew that 1939 paper; I had discovered it years before while still a graduate student, and I had been impressed by it. The paper dealt with a new class of distributions Neyman had discovered, which he called the "contagious distributions." The problem posed in the paper began with trying to model the appearance of insect grubs in soil. The female insect flew about the field, laden with eggs, then chose a spot at random to lay the eggs. Once the eggs were laid, the grubs hatched and crawled outward from that spot. A sample of soil is taken from the field. What is the probability distribution of the number of grubs found in that sample?

The contagious distributions describe such situations. They are derived in the 1939 paper with an apparently simple series of equations. This derivation seems obvious and natural. It is clear, when the reader gets to the end of the paper, that there is no other way to approach it, but this is clear only after reading Neyman. Since that 1939 paper, Neyman's contagious distributions have been found to fit a great many situations in medical research, in metallurgy, in meteorology, in toxicology, and (as described by Neyman in his Pfizer Colloquium paper) in dealing with the distribution of galaxies in the universe.

After he finished his talk, Neyman sat back to listen to the panel of discussants. All the other members of the panel were prominent statisticians who had been too busy to read his paper in advance. They looked upon the Pfizer Colloquium as a recognition of honor for Neyman. Their "discussions" consisted of comments about Neyman's career and past accomplishments. I had come onto the panel as a last-minute replacement and could not refer to my (nonexistent) previous experiences with Neyman. My comments were directed to Neyman's presentation that day, as he had asked. In particular, I told how I had discovered the 1939 paper years before and revisited it in anticipation of this session. I described the paper, as best I could, showing enthusiasm when I came to the clever way Neyman had developed the meaning of the parameters of the distribution.

Neyman was clearly delighted with my comments. Afterward, we had an exciting discussion about the contagious distributions and their uses. A few weeks later, a large package arrived in the mail. It was a copy of *A Selection of Early Statistical Papers of J. Neyman*, published by the University of California Press. On the inside cover was the inscription: "To Dr. David Salsburg, with hearty thanks for his interesting comments on my presentation of April 30, 1974. J. Neyman."

I treasure this book both for the inscription and the set of beautiful, well-written papers in it. I have since had the opportunity to talk with many of Neyman's students and coworkers. The friendly, interesting, and interested man I met in 1974 was the man that they knew and admired.

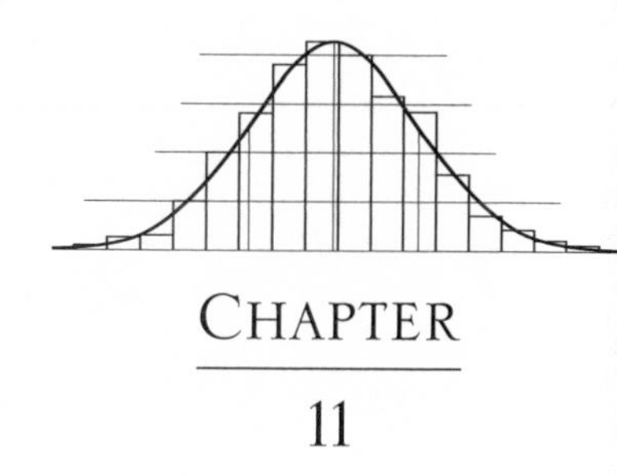

CHAPTER

11

HYPOTHESIS TESTING

At the start of their collaboration, Egon Pearson asked Jerzy Neyman how he could be sure that a set of data was normally distributed if he failed to find a significant p-value when testing for normality. Their collaboration started with this question, but Pearson's initial question opened the door to a much broader one. What does it mean to have a nonsignificant result in a significance test? Can we conclude that a hypothesis is true if we have failed to refute it?

R. A. Fisher had addressed that question in an indirect way. Fisher would take large p-values (and a failure to find significance) as indicating that the data were inadequate to decide. To Fisher, there was never a presumption that a failure to find significance meant that the tested hypothesis was true. To quote him:

> For the logical fallacy of believing that a hypothesis has been proved to be true, merely because it is not contradicted by the available facts, has no more right to insinuate itself in statistical than in

> other kinds of scientific reasoning. . . . It would, therefore, add greatly to the clarity with which the tests of significance are regarded if it were generally understood that tests of significance, when used accurately, are capable of rejecting or invalidating hypotheses, in so far as they are contradicted by the data: but that they are never capable of establishing them as certainly true. . . .

Karl Pearson had often used his chi square goodness of fit test to "prove" that data followed particular distributions. Fisher had introduced more rigor into mathematical statistics, and Karl Pearson's methods were no longer acceptable. The question still remained. It was necessary to assume that the data fit a particular distribution, in order to know which parameters to estimate and determine how those parameters relate to the scientific question at hand. The statisticians were frequently tempted to use significance tests to prove that.

In their correspondence, Egon Pearson and Jerzy Neyman explored several paradoxes that emerged from significance testing, cases where unthinking use of a significance test would reject a hypothesis that was obviously true. Fisher never fell into those paradoxes, because it would have been obvious to him that the significance tests were being applied incorrectly. Neyman asked what criteria were being used to decide when a significance test was applied correctly. Gradually, between their letters, with visits that Neyman made to England during the summers and Pearson's visits to Poland, the basic ideas of hypothesis testing emerged.[1]

A simplified version of the Neyman-Pearson formulation of hypothesis testing can now be found in all elementary statistics text-

[1]Throughout this chapter, I attribute the essential mathematical ideas to Neyman. This is because Neyman was responsible for the polished final formulation and for the careful mathematical development behind it. However, correspondence between Egon Pearson and William Sealy Gosset, which began six months before Pearson met Neyman, indicates that Pearson was already thinking about alternative hypotheses and different types of errors and that Gosset may have first suggested the idea. In spite of the fact that his was the initial input, Pearson acknowledged that Neyman provided the mathematical foundations for his own "loose ideas."

books. It has a simple structure. I have found that it is easy for most first-year students to understand. Since it has been codified, this version of the formulation is exact and didactic. This is how it must be done, the texts imply, and this is the only way it can be done. This rigid approach to hypothesis testing has been accepted by regulatory agencies like the U.S. Food and Drug Administration and the Environmental Protection Agency, and it is taught in medical schools to future medical researchers. It has also wormed its way into legal proceedings when dealing with certain types of discrimination cases.

When the Neyman-Pearson formulation is taught in this rigid, simplified version of what Neyman developed, it distorts his discoveries by concentrating on the wrong aspects of the formulation. Neyman's major discovery was that significance testing made no sense unless there were at least two possible hypotheses. That is, you could not test whether data fit a normal distribution unless there was some other distribution or set of distributions that you believed it would fit. The choice of these alternative hypotheses dictates the way in which the significance test is run. The probability of detecting that alternative, if it is true, he called the "power" of the test. In mathematics, clarity of thought is developed by giving clear, well-defined names to specific concepts. In order to distinguish between the hypothesis being used to compute Fisher's p-value and the other possible hypothesis or hypotheses, Neyman and Pearson called the hypothesis being tested the "null hypothesis" and the other hypotheses the "alternative." In their formulation, the p-value is calculated for testing the null hypothesis but the power refers to how this p-value will behave if the alternative is, in fact, true.

This led Neyman to two conclusions. One was that the power of a test was a measure of how good the test was. The more powerful of two tests was the better one to use. The second conclusion was that the set of alternatives cannot be too large. The analyst cannot say that the data come from a normal distribution (the null hypothesis) or that they come from any other possible distribution. That is too wide a set of alternatives, and no test can be powerful against all possible alternatives.

In 1956, L. J. Savage and Raj Raghu Bahadur at the University of Chicago showed that the class of alternatives does not have to be very wide for hypothesis testing to fail. They constructed a relatively small set of alternative hypotheses against which no test had any power. During the 1950s, Neyman developed the idea of restricted hypothesis tests, where the set of alternative hypotheses is very narrowly defined. He showed that such tests are more powerful than ones dealing with more inclusive sets of hypotheses.

In many situations, hypothesis tests are used against a null hypothesis that is a straw man. For instance, when two drugs are being compared in a clinical trial, the null hypothesis to be tested is that the two drugs produce the same effect. However, if that were true, then the study would never have been run. The null hypothesis that the two treatments are the same is a straw man, meant to be knocked down by the results of the study. So, following Neyman, the design of the study should be aimed at maximizing the power of the resulting data to knock down that straw man and show how the drugs differ in effect.

WHAT IS PROBABILITY?

Unfortunately, to develop a mathematical approach to hypothesis testing that was internally consistent, Neyman had to deal with a problem that Fisher had swept under the rug. This is a problem that continues to plague hypothesis testing, in spite of Neyman's neat, purely mathematical solution. It is a problem in the application of statistical methods to science in general. In its more general form, it can be summed up in the question: What is meant by probability in real life?

The mathematical formulations of statistics can be used to compute probabilities. Those probabilities enable us to apply statistical methods to scientific problems. In terms of the mathematics used, probability is well defined. How does this abstract concept connect to reality? How is the scientist to interpret the probability statements of statistical analyses when trying to decide what is true and what is not? In the final chapter of this book I shall examine the general problem and the attempts that have been made to answer these questions. For the moment, however, we will examine the specific circumstances that forced Neyman to find his version of an answer.

Recall that Fisher's use of a significance test produced a number Fisher called the p-value. This is a calculated probability, a probability associated with the observed data under the assumption that the null hypothesis is true. For instance, suppose we wish to test a new drug for the prevention of a recurrence of breast cancer in patients who have had mastectomies, comparing it to a placebo. The null hypothesis, the straw man, is that the drug is no better than the placebo. Suppose that after five years, 50 percent of the women on placebos have had a recurrence and none of the women on the new drug have. Does this prove that the new drug "works"? The answer, of course, depends upon how many patients that 50 percent represents.

If the study included only four women in each group, that means we had eight patients, two of whom had a recurrence. Suppose we take any group of eight people, tag two of them, and divide the eight at random into two groups of four. The probability that both of the tagged people will fall into one of the groups is around .30. If there were only four women in each group, the fact that all the recurrences fell in the placebo group is not significant. If the study included 500 women in each group, it would be highly unlikely that all 250 with recurrences were on the placebo, unless the drug was working. The probability that all 250 would fall in one group if the drug was no better than the placebo is the p-value, which calculates to be less than .0001.

The p-value is a probability, and this is how it is computed. Since it is used to show that the hypothesis under which it is calculated is false, what does it really mean? It is a theoretical probability associated with the observations under conditions that are most likely false. It has nothing to do with reality. It is an indirect measurement of plausibility. It is not the probability that we would be wrong to say the drug works. It is not the probability of any kind of error. It is not the probability that a patient will do as well on the placebo as on the drug. But, to determine which tests are better than others, Neyman had to find a way to put hypothesis testing within a framework wherein probabilities associated with the decisions made from the test could be calculated. He needed to connect the p-values of the hypothesis test to real life.

THE FREQUENTIST DEFINITION OF PROBABILITY

In 1872, John Venn, the British philosopher, had proposed a formulation of mathematical probability that would make sense in real life. He turned a major theorem of probability on its head. This is the law of large numbers, which says that if some event has a given probability (like throwing a single die and having it land with the six side up) and if we run identical trials over and over again, the proportion of times that event occurs will get closer and closer to the probability.

Venn said the probability associated with a given event is the long-run proportion of times the event occurs. In Venn's proposal, the mathematical theory of probability did not imply the law of large numbers; the law of large numbers implied probability. This is the frequentist definition of probability. In 1921, John Maynard Keynes[2] demolished this as a useful or even meaningful interpretation, showing that it has fundamental inconsistencies that make

[2]There is a kind of misonomy involved with Keynes. Most people would think of him as an economist, the founder of the Keynesian school of economics, dealing with the ways in which government manipulation of monetary policy can influence the course of the economy. However, Keynes had his Ph.D. in philosophy; and his Ph.D. dissertation, published in 1921 as A *Treatise on Probability*, is a major monument in the development of the philosophical foundations behind the use of mathematical statistics. In future chapters, we will have occasion to quote Keynes. It will be from Keynes the probabilist, and not Keynes the economist, that we will be quoting.

it impossible to apply the frequentist definition in most cases where probability is invoked.

When it came to structuring hypothesis tests in a formal mathematical way, Neyman fell back upon Venn's frequentist definition. Neyman used this to justify his interpretation of the p-value in a hypothesis test. In the Neyman-Pearson formulation, the scientist sets a fixed number, such as .05, and rejects the null hypothesis whenever the significance test p-value is less than or equal to .05. This way, in the long run, the scientist will reject a true null hypothesis exactly 5 percent of the time. The way hypothesis testing is now taught, Neyman's invocation of the frequentist approach is emphasized. It is too easy to view the Neyman-Pearson formulation of hypothesis testing as a part of the frequentist approach to probability and to ignore the more important insights that Neyman provided about the need for a well-defined set of alternative hypotheses against which to test the straw man of the null hypothesis.

Fisher misunderstood Neyman's insights. He concentrated on the definition of significance level, missing the important ideas of power and the need to define the class of alternatives. In criticism of Neyman, he wrote:

> Neyman, thinking he was correcting and improving my own early work on tests of significance, as a means to the "improvement of natural knowledge," in fact reinterpreted them in terms of that technological and commercial apparatus which is known as an acceptance procedure. Now, acceptance procedures are of great importance in the modern world. When a large concern like the Royal Navy receives material from an engineering firm it is, I suppose, subjected to sufficiently careful inspection and testing to reduce the frequency of the acceptance of faulty or defective consignments. . . . But, the logical differences between such an operation and the work of scientific discovery by physical or

> biological experimentation seem to me so wide that the analogy between them is not helpful, and the identification of the two sorts of operations is decidedly misleading.

In spite of these distortions of Neyman's basic ideas, hypothesis testing has become the most widely used statistical tool in scientific research. The exquisite mathematics of Jerzy Neyman have now become an idée fixe in many parts of science. Most scientific journals require that the authors of articles include hypothesis testing in their data analyses. It has extended beyond the scientific journals. Drug regulatory authorities in the United States, Canada, and Europe require the use of hypothesis tests in submissions. Courts of law have accepted hypothesis testing as an appropriate method of proof and allow plaintiffs to use it to show employment discrimination. It permeates all branches of statistical science.

The climb of the Neyman-Pearson formulation to the pinnacle of statistics did not go unchallenged. Fisher attacked it from its inception and continued to attack it for the rest of his life. In 1955, he published a paper entitled "Statistical Methods and Scientific Induction" in the *Journal of the Royal Statistical Society*, and he expanded on this with his last book, *Statistical Methods and Scientific Inference*. In the late 1960s, David Cox, soon to be the editor of *Biometrika*, published a trenchant analysis of how hypothesis tests are actually used in science, showing that Neyman's frequentist interpretation was inappropriate to what is actually done. In the 1980s, W. Edwards Deming attacked the entire idea of hypothesis testing as nonsensical. (We shall come back to Deming's influence on statistics in chapter 24.) Year after year, articles continue to appear in the statistical literature that find new faults with the Neyman-Pearson formulation as frozen in the textbooks.

Neyman himself took no part in the canonization of the Neyman-Pearson formulation of hypothesis testing. As early as 1935, in an article he published (in French) in the *Bulletin de la Société Mathé-*

matique de France, he raised serious doubts about whether optimum hypothesis tests could be found. In his later papers, Neyman seldom made use of hypothesis tests directly. His statistical approaches usually involved deriving probability distributions from theoretical principles and then estimating the parameters from the data.

Others picked up the ideas behind the Neyman-Pearson formulation and developed them. During World War II, Abraham Wald expanded on Neyman's use of Venn's frequentist definitions to develop the field of statistical decision theory. Eric Lehmann produced alternative criteria for good tests and then, in 1959, wrote a definitive textbook on the subject of hypothesis testing, which remains the most complete description of Neyman-Pearson hypothesis testing in the literature.

Just before Hitler invaded Poland and dropped a curtain of evil upon continental Europe, Neyman came to the United States, where he started a statistics program at the University of California at Berkeley. He remained there until his death in 1981, having created one of the most important academic statistics departments in the world. He brought to his department some of the major figures in the field. He also drew from obscurity others who went on to great achievements. For example, David Blackwell was working alone at Howard University, isolated from other mathematical statisticians. Because of his race, he had been unable to get an appointment at "White" schools, in spite of his great potential; Neyman invited Blackwell to Berkeley. Neyman also brought in a graduate student who had come from an illiterate French peasant family; Lucien Le Cam went on to become one of the world's leading probabilists.

Neyman was always attentive to his students and fellow faculty members. They describe the pleasures of the afternoon departmental teas, which Neyman presided over with a courtly graciousness. He would gently prod someone, student or faculty, to describe some recent research and then genially work his way around the room, getting comments and aiding the discussion. He would end many teas by lifting his cup and toasting, "To the ladies!" He was especially good to "the ladies," encouraging and furthering the careers of women. Prominent among his female protégées were Dr. Elizabeth Scott, who worked with Neyman and was a coauthor on papers ranging from astronomy to carcinogenesis to zoology, and Dr. Evelyn Fix, who made major contributions to epidemiology.

Until R. A. Fisher died in 1962, Neyman was under constant attack by this acerbic genius. Everything Neyman did was grist for Fisher's criticism. If Neyman succeeded in showing a proof of some obscure Fisherian statement, Fisher attacked him for misunderstanding what he had written. If Neyman expanded on a Fisherian idea, Fisher attacked him for taking the theory down a useless path. Neyman never responded in kind, either in print or, if we are to believe those who worked with him, in private.

In an interview toward the end of his life, Neyman described a time in the 1950s when he was about to present a paper in French at an international meeting. As he went to the podium, he realized that Fisher was in the audience. While presenting the paper, he steeled himself for the attacks he knew would come. He knew that Fisher would pounce upon some unimportant minor aspect of the paper and tear it and Neyman to pieces. Neyman finished and waited for questions from the audience. A few came. But Fisher never stirred, never said a word. Later, Neyman discovered that Fisher could not speak French.

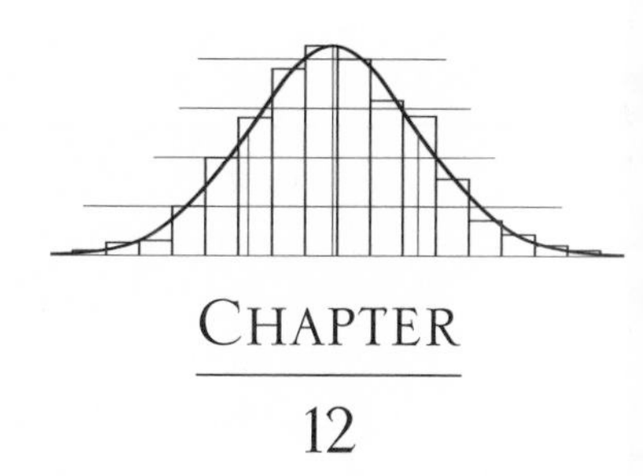

CHAPTER

12

THE CONFIDENCE TRICK

When the AIDS epidemic appeared in the 1980s, a number of questions needed to be answered. Once the infective agent, HIV (for human immunodeficiency virus), was identified, health officials needed to know how many infected people there were, in order to plan on the resources that would be needed to meet this epidemic. Fortunately, mathematical models of epidemiology[1] developed over the previous twenty to thirty years could be applied here.

The modern scientific view of epidemic disease is that individual patients are exposed, some of them become infected, and after a period of time called the "latency," many of those who became infected show symptoms of the disease. Once infected, a person is a potential source of exposure to others who

[1]Epidemiology is an allied field of statistics in which statistical models are used to examine patterns of human health. In its simplest form, epidemiology provides tabulations of vital statistics along with simple estimates of the parameters of their distributions. In its more complicated forms, epidemiology makes use of advanced theories of statistics to examine and predict the course of epidemic diseases.

are not yet infected. There is no way we can predict which person will be exposed, or infected, or will infect others. Instead, we deal with probability distributions and estimate the parameters of those distributions.

One of the parameters is the mean latency time—the average time from infection to symptoms. This was a particularly important parameter to public health officials for the AIDS epidemic. They had no way of knowing how many persons were infected and how many would eventually come down with the disease, but if they knew the mean latency time, they could combine that with the count of people who had the disease and estimate the number infected. Furthermore, due to an unusual circumstance in the infection pattern of AIDS, they had a group of patients about whom they knew both the time of infection and the time the disease appeared. A small number of hemophiliacs had been exposed to HIV through contaminated blood products, and they provided the data to estimate the parameter of mean latency time.

How good was this estimate? The epidemiologists could point out that they used the best estimates, in R. A. Fisher's sense. Their estimates were consistent and had maximum efficiency. They could even correct for possible bias and claim that their estimates were unbiased. But, as pointed out in earlier chapters, there is no way of knowing if a specific estimate is correct.

If we cannot say the estimate is exactly correct, is there some way we can say how close the estimate is to the true value of the parameter? The answer to that question lies in the use of interval estimates. A point estimate is a single number. For instance, we might use the data from hemophiliac studies to estimate that the mean latency time was 5.7 years. An interval estimate would state that the mean latency time lies between 3.7 and 12.4 years. Very often, it is adequate to have an interval estimate, since the public policies that would be required are about the same for both ends of the interval estimate. Sometimes, the interval estimate is too wide, and different public policies would be required for the minimum value than for the maximum value. The conclusion that can be drawn from too wide an interval is that the information available is not adequate to make a decision and that more information should be sought, perhaps by enlarging the scope of the investigation or engaging in another series of experiments.

For instance, if the mean latency time for AIDS is as high as 12.4 years, then approximately one-fifth of infected patients will survive for 20 years or more after being infected, before they get AIDS. If the mean latency is 3.7 years, almost every patient will get AIDS within 20 years. These two results are too disparate to lead to a single best public policy, and more information would be useful.

In the late 1980s, the National Academy of Sciences convened a committee of some of the country's leading scientists to consider the possibility that fluorocarbons used in aerosol sprays were destroying the ozone layer in the upper atmosphere, which protects the Earth from harmful ultraviolet radiation. Instead of answering the question with a yes or no, the committee (whose chairman, John Tukey, is the subject of chapter 22 of this book) decided to model the effect of fluorocarbons in terms of a probability distribution. They then computed an interval estimate of the mean change in ozone per year. It turned out that, even using the small amount of data available, the lower end of that interval indicated a sufficient annual decrease in ozone to pose a serious threat to human life within fifty years.

Interval estimates now permeate almost all statistical analyses. When a public opinion poll claims that 44 percent of the populace think the president is doing a good job, there is usually a footnote stating this figure has "an error of plus or minus 3 percent." What that means is that 44 percent of the people surveyed thought the president is doing a good job. Since this was a random survey, the parameter being sought is the percentage of all the people who think that way. Because of the small size of the sample, a reasonable guess is that the parameter runs from 41 percent (44 percent minus 3 percent) to 47 percent (44 percent plus 3 percent).

Jerzy Neyman, 1894–1981

How does one compute an interval estimate? How does one interpret an interval estimate? Can we make a probability statement about an interval estimate? How sure are we that the true value of the parameter lies within the interval?

NEYMAN'S SOLUTION

In 1934, Jerzy Neyman presented a talk before the Royal Statistical Society, entitled "On the Two Different Aspects of the Representative Method." His paper dealt with the analysis of sample surveys; it has the elegance of most of his work, deriving what appear to be simple mathematical expressions that are intuitively obvious (only after Neyman has derived them). The most important part of this paper is in an appendix, in which Neyman proposes a straightforward way to create an interval estimate and to determine how accurate that estimate is. Neyman called this new procedure "confidence intervals," and the ends of the confidence intervals he called "confidence bounds."

Professor G. M. Bowley was in the chair for the meeting and rose to propose a vote of thanks. He first discussed the main part of the paper for several paragraphs. Then he got to the appendix:

> I am not certain whether to ask for an explanation or to cast a doubt. It is suggested in the paper that the work is difficult to follow and I may be one of those who have been misled by it [later in this paragraph, he works out an example, showing that he clearly understood what Neyman was proposing]. I can only say that I have read it at the time it appeared and since, and I read Dr. Neyman's elucidation of it yesterday with great care. I am referring to Dr. Neyman's confidence limits. I am not at all sure that the "confidence" is not a "confidence trick."

Bowley then works out an example of Neyman's confidence interval and continues:

> Does that really take us any further? Do we know more than was known to Todhunter [a late-nineteenth-century probabilist]? Does it take us beyond Karl Pearson and Edgeworth [a leading figure in the early development of mathematical statistics]? Does it really lead us towards what we need—the chance that in the universe which we are sampling the proportion is within these certain limits? I think it does not. . . . I do not know that I have expressed my thoughts quite accurately . . . [this] is a difficulty I have felt since the method was first propounded. The statement of the theory is not convincing, and until I am convinced I am doubtful of its validity.

Bowley's problem with this procedure is one that has bedeviled the idea of confidence bounds since then. Clearly, the elegant four lines of calculus that Neyman used to derive his method are correct within the abstract mathematical theory of probability, and it does lead to the computation of a probability. However, it is not clear what that probability refers to. The data have been observed, the parameter is a fixed (if unknown) number, so the probability that the parameter takes on a specific value is either 100 percent if that was the value, or 0 if it was not. Yet, a 95 percent confidence interval deals with 95 percent probability. The probability of what? Neyman finessed this question by calling his creation a confidence interval and avoiding the use of the word probability. Bowley and others easily saw through this transparent ploy.

R. A. Fisher was also among the discussants, but he missed this point. His discussion was a rambling and confused collection of references to things that Neyman did not even include in his paper. This is because Fisher was in the midst of confusion over the calculation of interval estimates. In his comments, he referred to "fiducial probability," a phrase that does not appear in Neyman's paper. Fisher had long been struggling with this very problem—how to determine the degree of uncertainty associated with an interval estimate of a parameter. Fisher was working at the problem from a complicated angle somewhat related to his likelihood function. As he quickly proved, this way of looking at the formula did not meet the requirements of a probability distribution. Fisher called this function a "fiducial distribution," but then he violated his own insights by applying the same mathematics one might apply to a proper probability distribution. The result, Fisher hoped, would be a set of values that was reasonable for the parameter, in the face of the observed data.

This was exactly what Neyman produced, and if the parameter was the mean of the normal distribution, both methods produced the same answers. From this, Fisher concluded that Neyman had stolen his idea of fiducial distribution and given it a different name.

Fisher never got far with his fiducial distributions, because the method broke down with other more complicated parameters, like the standard deviation. Neyman's method works with any type of parameter. Fisher never appeared to understand the difference between the two approaches, insisting to the end of his life that Neyman's confidence intervals were, at most, a generalization of his fiducial intervals. He was sure that Neyman's apparent generalization would break down when faced with a sufficiently complicated problem—just as his own fiducial intervals had.

Probability Versus Confidence Level

Neyman's procedure does not break down, regardless of how complicated the problem, which is one reason it is so widely used in statistical analyses. Neyman's real problem with confidence intervals was not the problem that Fisher anticipated. It was the problem that Bowley raised at the beginning of the discussion. What does probability mean in this context? In his answer, Neyman fell back on the frequentist definition of real-life probability. As he said here, and made clearer in a later paper on confidence intervals, the confidence interval has to be viewed not in terms of each conclusion but as a process. In the long run, the statistician who always computes 95 percent confidence intervals will find that the true value of the parameter lies within the computed interval 95 percent of the time. Note that, to Neyman, the probability associated with the confidence interval was not the probability that we are correct. It was the frequency of correct statements that a statistician who uses his method will make in the long run. It says nothing about how "accurate" the current estimate is.

As careful as Neyman was to define the concept, and as careful as were statisticians like Bowley to keep the concept of probability clear and uncontaminated, the general use of confidence intervals in science has led to much more sloppy thinking. It is not uncommon, for instance, for someone using a 95 percent confidence interval to state that he is "95 percent sure" that the parameter lies within that interval. In chapter 13, we will meet L. J. ("Jimmie") Savage and Bruno de Finetti and describe their work on personal probability, which justifies the use of statements like that. However, the calculation of the degree to which a person can be sure of something is different from the calculation of a confidence interval. The statistical literature has many articles where the bounds on a parameter that are derived following the methods of Savage and de Finetti are shown to be dramatically different from Neyman's confidence bounds derived from the same data.

In spite of the questions about the meaning of probability in this context, Neyman's confidence bounds have become the standard method of computing an interval estimate. Most scientists compute 90 percent or 95 percent confidence bounds and act as if they are sure that the interval contains the true value of the parameter.

No one talks or writes about "fiducial distributions" today. The idea died with Fisher. As he tried to make the idea work, Fisher produced a great deal of clever and important research. Some of that research has become mainstream; other parts remain in the incomplete state in which he left them.

In this research, Fisher came very close at times to stepping over the line into a branch of statistics he called "inverse probability." Each time he pulled away. The idea of inverse probability began with the Reverend Thomas Bayes, an amateur mathematician of the eighteenth century. Bayes maintained a correspondence with many of the leading scientists of his age and often posed complicated mathematical problems to them. One day, while fiddling with the standard mathematical formulas of probability, he combined two of them with simple algebra and discovered something that horrified him.

In the next chapter, we shall look at the Bayesian heresy and why Fisher refused to make use of inverse probability.

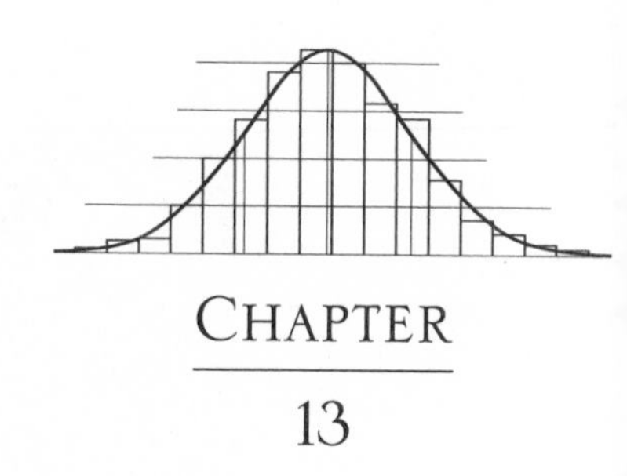

CHAPTER

13

THE BAYESIAN HERESY

The Serene Republic of Venice was a major power in the Mediterranean from the eighth through the early eighteenth centuries. At the height of its empire, Venice had control of much of the Adriatic coast and the islands of Crete and Cyprus, and had a monopoly over trade from the Orient to Europe. It was ruled by a group of noble families who maintained a kind of democracy among themselves. The titular head of state was the doge. From the founding of the republic in 697 until Venice was taken over by Austria in 1797, over 150 men served as doge, some for a year or less, one for as long as thirty-four years. Upon the death of the reigning doge, the republic engaged in an elaborate sequence of elections. From the senior members of the noble families, a small number would be chosen by lot as lectors. These lectors would choose additional members to join them at this first stage, and then a small number of this augmented group would be chosen by lot. This continued for several stages until a final group of lectors would choose the doge from among themselves.

The palace of the doge at the height of Venetian power

Early in the history of the republic, the lectors were chosen at each stage by preparing a group of wax balls, some of which had nothing in them and some of which had a slip of paper with the word *lector* on it. By the seventeenth century, the final stages were conducted using identically sized balls of gold and silver. When the Doge Rainieri Zeno died in 1268, there were thirty lectors at the second stage, and thirty balls of wax were prepared, nine of which had slips of paper in them. A small child was brought forward. He chose a ball from the basket and handed it to the first lector, who opened it and saw whether he would be a lector at the next stage. The child chose another ball and handed it to the second lector, who opened it, and so on.

Before the child chose the first ball, each member of the group had a probability of $9/30$ of becoming a lector for the next stage. If the first ball was empty, then each of the remaining members had a $^{9}/_{29}$ probability of being chosen. If the first ball contained a slip of paper, then each of the remaining members had an $^{8}/_{29}$ chance of being chosen. Once the second ball was chosen and displayed, the probability that the next member would be chosen lector would be similarly diminished or increased, depending upon the outcome of that draw. This would continue until all nine marked balls had been chosen. At that point, the chance that any of the remaining members would become a lector for the next stage dropped to zero.

This is an example of conditional probability. The probability that a given member would become a lector at the next stage depended upon the balls that had been chosen prior to his choice. John Maynard Keynes has pointed out that all probabilities are conditional. To use one of his examples, the probability that a book chosen at random from his library shelves will be bound in buckram is conditional on the books that are actually in his library and on how he will be making that choice "at random." The probability that a patient will have small-cell carcinoma of the lung is conditional on the smoking history of that patient. The p-value that is calculated to test the null hypothesis of no treatment effect from a controlled experiment is conditional on the design of the experiment. The important aspect of conditional probability is that the probability of some given event (e.g., that a particular set of numbers will win in a lottery) is different for different prior conditions.

The formulas that were developed during the eighteenth century to deal with conditional probability all depended upon the idea that the conditioning events occurred prior to the event being sought. In the latter part of that century, the Reverend Thomas Bayes was playing around with the formulas for conditional probability and made a startling discovery: The formulas had an inner symmetry.

Suppose we have two events that occur over a period of time, like shuffling a deck of cards and then dealing a poker hand of five. Let us call the events "before" and "after." It makes sense to talk about the probability of "after," conditional on "before." If we fail to shuffle the deck well, that will influence the probability of getting two aces in the poker hand. Bayes discovered that we can also calculate the probability of "before," conditional on "after." This made no sense. It would be like determining the probability that the deck of cards contained four aces, given that a poker hand has been dealt with two aces. Or the probability that a patient was a smoker, given that he had lung cancer. Or the probability that a state lottery was fair, given that someone named Charles A. Smith was the only winner.

Bayes put these calculations aside. They were found among his papers when he died, and were published posthumously. Bayes's theorem[1] has since come to bedevil the mathematics of statistical analysis. Far from failing to make sense, Bayes's inversion of conditional probability often makes a great deal of sense. When epidemiologists attempt to find the possible causes of a rare medical condition, like Reye's syndrome, they often use a case-control study. In such a study, a group of cases of the disease is assembled, and they are compared with a group of patients (the controls) who have not had this disease but who are similar in other respects to the patients with the disease. The epidemiologists calculate the probability of some prior treatment or condition, given that the control patients had the disease. This was how the effects of smoking on both heart disease and lung cancer were first discovered. The influence of thalidomide on birth defects was also deduced from a case-control study.

More important than the direct use of Bayes's theorem to invert conditional probability has been the use of the theorem for estimating parameters of distributions. There is a temptation to treat the parameters of a distribution as random themselves and to com-

---

[1]Stigler's law of misonomy comes into full bloom with this name. Bayes was far from being the first person to note the symmetry of conditional probability. The Bernoullis seem to have been aware of it. de Moivre makes reference to it. However, Bayes alone gets the credit (or, given Bayes's reluctance to publish, we might say Bayes gets the blame).

pute probabilities associated with those parameters. For instance, we might compare two treatments for cancer and want to conclude that "We are 95 percent sure that the five-year survival rate for treatment A is greater than the survival rate for treatment B." This can be done with one or two applications of Bayes's theorem.

Questions About "Inverse Probability"

For many years, the use of Bayes's theorem in this fashion was considered an inappropriate practice. There are serious questions about what probability means when used for parameters. After all, the whole basis of the Pearsonian revolution was that the measurements of science were no longer considered the things of interest. Rather, as Pearson showed, it was the probability distribution of those measurements, and the purpose of scientific investigation was to estimate the parameters whose (fixed but unknown) values controlled that distribution. If the parameters were considered random (and conditional on the observed measurements), this approach would cease to have such clear meaning.

During the early years of the twentieth century, statisticians were very careful to avoid "inverse probability," as it was called. In discussions before the Royal Statistical Society following one of his early papers, Fisher was accused of using inverse probability and stoutly defended himself against such a terrible charge. In the first paper on confidence intervals, Neyman seemed to be using inverse probability, but only as a mathematical device to get around a specific calculation. In his second paper, he showed how to reach the same result without Bayes's theorem. By the 1960s, the potential power and usefulness of such an approach began attracting more and more workers. This Bayesian heresy was becoming more and more respectable. By the end of the twentieth century, it had reached such a level of acceptability that over half the articles that appear in journals like the *Annals of Statistics* and *Biometrika* now make use of Bayesian methods. The application of Bayesian methods is still often suspect, especially in medical science.

One difficulty with explaining the Bayesian heresy is that there are a number of different methods of analysis and at least two different philosophical foundations to the use of these methods. It often seems as if entirely different ideas have been given the same label—Bayesian. In what follows, I will consider two specific formulations of the Bayesian heresy: the Bayesian hierarchal model, and personal probability.

THE BAYESIAN HIERARCHAL MODEL

In the early 1970s, statistical methods of textual analysis made great advances, starting with work by Frederick Mosteller and David Wallace, who used statistical methods to determine the authorship of the disputed *Federalist* papers. Leading up to the ratification of the new Constitution of the United States by the state of New York in 1787–1788, James Madison, Alexander Hamilton, and John Jay wrote a series of seventy articles supporting ratification. The articles were signed with pseudonyms. Early in the next century, Hamilton and Madison identified the papers each claimed to have written. They both claimed[2] twelve of the papers as their own.

In their statistical analysis of the disputed papers, Mosteller and Wallace identified several hundred English words that lacked "content." These are words like, *if, when, because, over, whilst, as, and.* The words are needed to give a sentence grammatical meaning, but they do not carry specific meaning, and their use is dependent primarily upon the way in which the author uses language. Of these hundreds of contentless words, they found about thirty where the two authors differed in the frequency of use in their other writings.

Madison, for example, used the word *upon* an average of 0.23 times in 1,000 words, and Hamilton used *upon* an average

[2]Actually, only Madison made the claim for himself. This was in response to a list of papers purported to have been written by Hamilton and released by his friends three years after his death.

of 3.24 times per 1,000. (Eleven of the twelve disputed papers do not use the word *upon* at all, and the other paper has an average of 1.1 uses per 1,000 words.) These average frequencies do not describe any specific collection of 1,000 words. The fact that they are not whole numbers means that they do not describe any observed sequence of words. They are, however, estimates of one of the parameters of the distribution of words in the writings of two different men.

The question in the disputed authorship for a given paper was: Do the patterns of usage of these words come from the probability distributions associated with Madison or from the probability distributions associated with Hamilton? These distributions have parameters, and the specific parameters that define Madison's or Hamilton's works differ. The parameters can only be estimated from their works, and these estimates could be wrong. The attempt to distinguish which distribution can be applied to a disputed paper is clouded with this uncertainty.

One way of getting an estimate of the level of uncertainty is to note that the exact values of these parameters for the two men are taken from a distribution that describes the parameters used by all educated people writing in English in North America in the late eighteenth century. For instance, Hamilton used the word *in* 24 times per 1,000 words. Madison used *in* 23 times per 1,000, and other contemporary writers tended to use *in* around 22 to 25 times per 1,000.

Subject to the patterns associated with the general usage of words at that time and place, the parameters for each man are random and have a probability distribution. In this way, the parameters that drive the usage of contentless words by Hamilton or Madison have, themselves, parameters, which we can call "hyperparameters." Using written works from other authors of the time and place, we can estimate the hyperparameters.

The English language is always changing from place to place and time to time. In twentieth-century English literature, for instance, the frequency of the use of the word *in* tends to be less than 20 per 1,000, indicating a slight shift in patterns of usage over the 200 years or more since the time of Hamilton and Madison. We can consider the hyperparameters that define the distribution of parameters in eighteenth-century North America as themselves having a probability distribution across all times and all places. We can use writings from other times and places, in addition to the writings from eighteenth-century North America, to estimate the parameters of these hyperparameters, which we can call "hyper-hyperparameters."

By repeated use of Bayes's theorem, we can determine the distribution of the parameters, then of the hyperparameters. In principle, we could extend this hierarchy further by finding the distribution of the hyper-hyperparameters in terms of hyper-hyper-hyperparameters, and so on. In this case, there is no obvious candidate for the generation of an additional level of uncertainty. Using the estimates of hyper- and hyper-hyperparameters, Mosteller and Wallace were able to measure the probability associated with the statement: Madison (or Hamilton) wrote this paper.

Hierarchal Bayesian models have been applied very successfully since the early 1980s to many difficult problems in engineering and biology. One such problem arises when the data seem to come from two or more distributions. The analyst proposes the existence of an unobserved variable that defines which distribution a given observation comes from. This identifying marker is a parameter, but it has a probability distribution (with hyperparameters) that can be incorporated into the likelihood function. Laird and Ware's EM algorithm is particularly suited to this type of problem.

This extensive use of Bayesian methods in the statistical literature is filled with confusion and disputation. Different methods producing different results can be proposed, and there are no clear criteria to determine which are correct. Traditionalists object to the use of Bayes's theorem in general, and Bayesians disagree about the details of their models. The situation is crying for another genius like R. A. Fisher to come and find a unifying principle with which to resolve these arguments. As we enter the twenty-first century, no such genius seems to have arisen. The problem remains as elusive as it did for the Reverend Thomas Bayes more than 200 years ago.

Personal Probability

Another Bayesian approach appears to be on a much more solid foundation. This is the concept of personal probability. The idea has been around since the initial work on probability by the Bernoullis in the seventeenth century. In fact, the very word *probability* was created to deal with the sense of personal uncertainty.

L. J. ("Jimmie") Savage and Bruno de Finetti developed much of the mathematics behind personal probability in the 1960s and 1970s. I attended a lecture at a statistics meeting at the University of North Carolina in the late 1960s in which Savage was propounding some of these ideas. There are no such things as proven scientific facts, Savage claimed. There are only statements, about which people who call themselves scientists associate a high probability. For instance, he said, most of the people listening to him at that talk would associate a high probability to the statement: "The world is round." However, if we were to take a census of the world's population, we would most likely find many peasants in the middle of China who would associate a low probability to that statement. At that moment, Savage had to stop talking because a group of students at the university went racing by outside the hall shouting, "Close it down! Strike, strike! Close it down!" They were protesting the war in Vietnam by calling for a student strike at the university. As they disappeared down the path and the tumult died away, Savage looked at the window and said, "And, you know, we may be the last generation that thinks the world is round."

There are different versions of personal probability. At one extreme is the Savage–de Finetti approach, which says that each person has his or her own unique set of probabilities. At the other extreme is Keynes's view that probability is the degree of belief that an educated person in a given culture can be expected to hold. In Keynes's view, all people in a given culture (Savage's "scientists" or "Chinese peasants") can agree on a general level of probability that holds for a given statement. Because this level of probability depends upon the culture and time, it is very possible that the appropriate level of probability is wrong in some absolute sense.

Savage and de Finetti proposed that each individual has a specific set of personal probabilities, and they described how such probabilities might be elicited via a technique known as the "standard gamble." In order for an entire culture to share a given set of probabilities, Keynes had to weaken the mathematical definition and refer to probability not so much as a precise number (like 67 percent) but more as a method of ordering ideas (the probability that it will rain tomorrow is greater than the probability that it will snow).

Regardless of how the concept of personal probability is defined exactly, the way in which Bayes's theorem is used in personal probability seems to match the way in which most people think. The Bayesian approach is to start with a prior set of probabilities in the mind of a given person. Next, that person observes or experiments and produces data. The data are then used to modify the prior probabilities, producing a posterior set of probabilities:

$$\text{prior probability} \rightarrow \text{data} \rightarrow \text{posterior probability}$$

Suppose the person wishes to determine whether all ravens are black. She starts with some prior knowledge about the probability that this is true. For instance, she may know nothing about ravens at first and starts with equipose, 50:50, that all ravens are black. The data consist of her observations of ravens. Suppose she sees a raven and observes that it is black, her posterior probability is increased. The next time she observes ravens, her new prior (the old posterior) is greater than 50 percent and is further increased by observing a new set of ravens, all of which are black.

On the other hand, a person can enter the process with a very strong prior, so strong that it may take massive amounts of data to overcome it. In the near disaster of American nuclear power at the Three Mile Island power plant in Pennsylvania in the 1980s, the operators of the reactor had a large board of dials and indicators to follow the progress of the reactor. Among these were warning lights, some of which had been faulty and presented false alarms in the past. The prior beliefs of the operators were such that any new warning light would be viewed as a false alarm. Even as the pattern of warning lights and associated dials produced a consistent picture of low water in the reactor, they continued to dismiss the evidence. Their prior was so strong that the data did not change the posterior very much.

Suppose there are only two possibilities, as was the case for a disputed *Federalist* paper: It was written either by Madison or by Hamilton. Then, the application of Bayes's theorem leads to a simple relationship between the prior odds and the posterior odds where the data can be summarized into something called the "Bayes factor." This is a mathematical calculation that characterizes the data without reference to the prior odds at all. With this in hand, the analyst can tell the readers to insert whatever prior odds they wish, multiply it by the computed Bayes factor, and compute the posterior odds. Mosteller and Wallace did this for each of the twelve disputed *Federalist* papers.

They also ran two non-Bayesian analyses of the frequency of contentless words. This gave them four methods for determining the authorship of the disputed papers: the hierarchal Bayes model, the computed Bayes factor, and the two non-Bayesian analyses. How did the results come out? All twelve were overwhelmingly awarded to Madison. In fact, using the computed Bayes factors, for some of the papers the reader would have to have had prior odds greater than 100,000 to 1 in favor of Hamilton in order to produce posterior odds of 50:50.

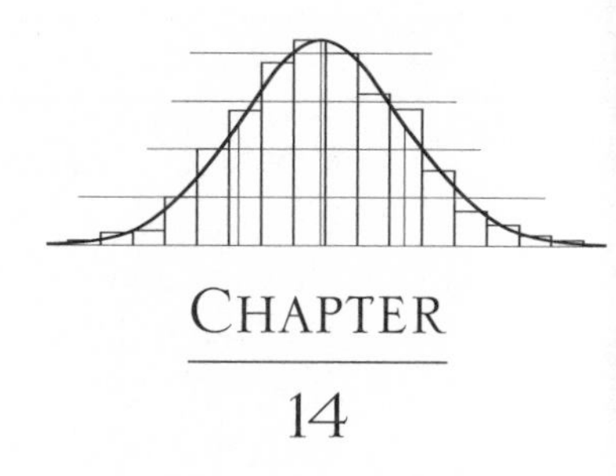

CHAPTER

14

THE MOZART OF MATHEMATICS

R. A. Fisher was not the only presiding genius when it came to the development of statistical methods in the twentieth century. Andrei Nikolaevich Kolmogorov, who was thirteen years younger than Fisher and died at age eighty-five in 1987, left an imprint on mathematical statistics and probability theory that built on some of Fisher's work but that exceeds Fisher's in the depth and detail of the mathematics.

But perhaps as important as his contributions to science was the impact this remarkable man had on those who knew him. His student Albert N. Shiryaev wrote in 1991:

> Andrei Nikolaevich Kolmogorov belonged to a select group of people who left one with the feeling of having touched someone unusual, someone great and extraordinary, the feeling of having met a wonder. Everything about Kolmogorov was unusual: his entire life, his school and college years, his pioneering discoveries in . . .

> mathematics . . . meteorology, hydrodynamics, history, linguistics and pedagogy. His interests were unusually diverse, including music, architecture, poetry, and travel. His erudition was unusual; it seemed as if he had an educated opinion about everything. . . . One's feeling after meeting Kolmogorov, after a simple conversation with him, [was] unusual. One sensed that he had a continuously intensive brain activity.

Kolmogorov was born in 1903 while his mother was on a journey from the Crimea to her home in the village of Tunoshna in southern Russia; she died in childbirth. One of his biographers delicately states: "He was the son of parents not formally married." Mariya Yakovlevna Kolmogorova was returning home in the final stages of pregnancy, having been abandoned by her boyfriend. In the throes of birth pangs, she was taken off the train at the town of Tambov. There she gave birth and then died alone in a strange town. Only her infant son came back to Tunoshna. His mother's maiden sisters raised him. One of them, Vera Yakovlevna, became his surrogate mother. His aunts ran a small school for young Andrei and his friends in their village. They printed a house magazine, *Spring Swallows*, in which they published the first of his literary efforts. At age five, he made his first mathematical discovery (which appeared in *Spring Swallows*). He found that the sum of the first **k** odd numbers was equal to the square of **k**. As he grew older, he would pose problems to his classmates, and the problems and their solutions were printed in *Spring Swallows*. One example of these problems was: In how many ways is it possible to sew on a button with four holes?

At fourteen, Kolmogorov learned higher mathematics from an encyclopedia, filling in the missing proofs. In secondary school, he frustrated his young physics teacher by creating plans for a series of perpetual motion machines. The plans were so clever that the teacher was unable to discover the errors (which Kolmogorov had carefully concealed). He decided to take the final examinations for his secondary school a year early. He informed his teachers, was asked to return after lunch, and went for a walk. When he returned, the examination board gave him his certificate without testing him. He recalled to Shiryaev that this was one of the greatest disappointments of his life; he had looked forward to the intellectual challenge.

Kolmogorov arrived in Moscow in 1920, at age seventeen, to attend the university. He entered as a student in mathematics but went to lectures in other fields like metallurgy, and took part in a seminar in Russian history. As a part of that seminar, he presented his first piece of research to be published, an examination of landholding in Novgorod in the fifteenth and sixteenth centuries. His professor criticized the paper because he did not think Kolmogorov had provided enough proof of his thesis. An archeological expedition to the region years later confirmed Kolmogorov's conjectures.

As an undergraduate at Moscow State University, he worked part-time teaching in a secondary school and joined in a large number of extracurricular activities. He went on to graduate studies in mathematics at Moscow. There was a basic set of fourteen courses required by the department. The students had the choice of taking a final exam in a given course or presenting an original paper. Few students attempted more than one paper. Kolmogorov never took an exam. He prepared fourteen papers with brilliant, original results in each one. "One of these results turned out to be false," he recalled, "but I realized it only later."

Kolmogorov, the brilliant mathematician, was known to the scientists of the West through a series of remarkable papers and books published in German journals. He was even allowed to attend a few mathematics conferences in Germany and Scandinavia during the 1930s. Kolmogorov, the man, disappeared behind Stalin's iron curtain during and after World War II. In 1938, he had published a paper that established the basic theorems for smoothing and predicting stationary stochastic processes. (This work will be described later in the chapter.) An interesting comment on the secrecy of war efforts comes from Norbert Wiener who, at the

A. N. Kolmogorov with students

Massachusetts Institute of Technology, worked on applications of these methods to military problems during and after the war. These results were considered so important to America's Cold War efforts that Wiener's work was declared top secret. But all of it, Wiener insisted, could have been deduced from Kolmogorov's early paper. During World War II, Kolmogorov was busy developing applications of the theory for the Soviet war effort. With the simple modesty that marked so much of his accomplishment, he attributed the basic ideas to R. A. Fisher, who had used similar methods in his genetics work.

KOLMOGOROV, THE MAN

When Stalin died in 1953, the iron ring of suspicion began to open. Kolmogorov, the man, emerged to participate in international meetings and to organize such meetings in Russia. The rest of the mathematical world now came to know him. He was an eager, friendly, open, and humorous man, with a wide range of interests and a love of teaching. His keen mind was always playing with what he was hearing. I have in front of me a photograph of Kolmogorov in the audience at a lecture being given by the British statistician David Kendall in Tbilisi in 1963. Kolmogorov's glasses are perched on the end of his nose. He is leaning forward, eagerly following the discussion. You can feel the vibrancy of his personality glowing in the midst of the others sitting around him.

Some of Kolmogorov's favorite activities were teaching and organizing classes at a school for gifted children in Moscow. He enjoyed introducing the children to literature and music. He took them on hikes and expeditions. He felt that each child should have a "broad and natural development of the whole personality," David Kendall wrote. "It did not worry him if they did not become mathematicians. Whatever profession they ultimately followed, he was content if their outlook remained broad and their curiosity unstifled."

Kolmogorov was married in 1942 to Anna Dmitrievna Egorova. He and his wife lived in a loving marriage into their eighties. He was an avid hiker and skier, and in his seventies he would take groups of young people hiking the trails of his favorite mountains, discussing mathematics, literature, music, and life in general. In 1971, he joined a scientific expedition exploring the oceans on the research vessel the *Dmitri Mendeleev*. His contemporaries were constantly amazed at the things that interested him and the knowledge he had. Upon meeting Pope John Paul II, he discussed skiing with this athletic pope, and then pointed out that during the nineteenth century, fat popes had alternated with thin popes and that Pope John Paul II was the 264th pope. It seems that one of his interests was the history of the Roman Catholic Church. He gave lectures on the statistical textual analysis of Russian poetry and could quote long selections from Pushkin by heart.

In 1953, a session was organized at Moscow State University to celebrate Kolmogorov's fiftieth birthday. Pavel Aleksandrov, professor emeritus, one of the speakers, said:

> Kolmogorov belongs to a group of mathematicians whose every single work in any area leads to a complete reevaluation. It is difficult to find a mathematician in

> recent years with not only such broad interests, but also such an influence on mathematics. . . . Hardy [a prominent British mathematician] took him for a specialist in trigonometric series, and von Karman [a post–World War II German physicist] took him for a specialist in mechanics. Gödel [a mathematical-philosophical theoretician] once said that the gist of human genius is the longevity of one's youth. Youth has several traits, one of which is excitement. Excitement with mathematics is one of the features of Kolmogorov's genius. Kolmogorov's excitement is in his creative work, in his articles in the *Large Soviet Encyclopedia*, in his development of the Ph.D. program. And this is just one side of him. The other is his dedicated labor.

And what were the results of this dedicated labor? It would be easier to list the fields of mathematics, physics, biology, and philosophy in which Kolmogorov did *not* have a major influence than it would be to list the areas in which his imprint was felt. In 1941, he founded the modern mathematical approach to turbulent fluid-flow. In 1954, he examined the gravitational interaction among the planets and found a way to model the "non-integrable" aspects that had defied mathematical analysis for over one hundred years.

KOLMOGOROV'S WORK IN MATHEMATICAL STATISTICS

Regarding the statistical revolution, Kolmogorov solved two of the most pressing theoretical problems. Before he died, he was close to the solution of a deep mathematical-philosophical problem that eats at the heart of statistical methods. The two pressing problems were

1. What are the true mathematical foundations of probability?

2. What can be done with data that are collected over time, like the vibrations of the earth following an earthquake (or an underground nuclear explosion)?

When Kolmogorov began examining the first question, probability was somewhat in ill repute among theoretical mathematicians. This was because the mathematical techniques of calculating probabilities had been developed during the eighteenth century as clever counting methods (viz., In how many ways can three sets of five cards be drawn from a standard deck so that only one hand is a winner?). These clever counting methods seemed to have no single underlying theoretical structure. They were almost all ad hoc, created to meet specific needs.

For most people, having a method for solving a problem would be adequate, but to late-nineteenth- and twentieth-century mathematicians a solid, rigorous underlying theory was necessary to make sure there would be no errors in those solutions. The ad hoc methods of the eighteenth-century mathematicians had worked but had also led to difficult paradoxes when applied incorrectly. The major work of early-twentieth-century mathematics involved putting those ad hoc methods on solid, rigorous mathematical foundations. The major reason why Henri Lebesgue's work was important (this was the Lebesgue who so impressed Jerzy Neyman with his mathematics but who turned out to be a discourteous boor when Neyman met him) was that it put the ad hoc methods of integral calculus on a solid foundation. As long as probability theory remained an incomplete seventeenth- and eighteenth-century invention, mathematicians of the twentieth century treated it as something of lesser value (and included statistical methods in that judgment).

Kolmogorov thought about the nature of probability calculations and finally realized that finding the probability of an event was exactly like finding the area of an irregular shape. He adapted the newly emerging mathematics of measure theory to the calculations of probabilities. With these tools, Kolmogorov was able to identify a small set of axioms upon which he could construct the entire body of probability theory. This is Kolmogorov's "axiomization of probability theory." It is taught today as the only way to view probability. It settles forever all questions about the validity of those calculations.

Having solved the problem of probability theory, Kolmogorov attacked the next major problem of statistical methods (in between teaching gifted children, organizing seminars, running a department of mathematics, solving problems in mechanics and astronomy, and living life to its fullest). To make statistical calculations possible, R. A. Fisher and other statisticians had assumed that all data are independent. They looked on a sequence of measurements as if it had been generated by tossing dice. Since the dice do not remember their previous configuration, each new number was completely independent of the previous ones.

Most data are not independent of each other. The first example Fisher used in *Statistical Methods for Research Workers* was the weekly weights of his newborn son. Clearly, if the child gained an unusual amount of weight one week, the next week's weight would reflect that, or if the child got sick and failed to gain one week, the next week's weight would reflect that. It is difficult to think of any sequence of data collected over time in real-life situations where successive observations are truly independent.

In the third of his "Studies in Crop Variation" (the massive paper that H. Fairfield Smith introduced me to), Fisher dealt with a series of wheat harvest measurements taken in successive years and rainfall measurements taken on successive days. He attacked the problem by creating a set of complicated parameters to account for the fact that data collected over time are not independent. He found a limited range of solutions that depended upon simplifying assumptions that may not be true. Fisher was unable to proceed much further, and no one followed up on his work.

No one, that is, until Kolmogorov. He called a sequence of numbers collected over time with the successive values related to previous ones a "stochastic process." Kolmogorov's pioneering papers (published just before the beginning of World War II) laid the foundations for further work by Norbert Wiener in the United States, George Box in England, and Kolmogorov's students in Russia. Because of Kolmogorov's ideas, it is now possible to examine records taken over time and come to highly specific conclusions. The waves lapping against a beach in California have been used to locate a storm in the Indian Ocean. Radio telescopes can distinguish between different sources (and perhaps, some day, intercept a message from intelligent life on another planet). It is possible to know whether a seismographic record is the result of an underground nuclear explosion or a natural earthquake. Engineering journals are filled with articles that make use of methods that have evolved out of Kolmogorov's work on stochastic processes.

What Is Probability in Real Life?

In the final years of his life, Kolmogorov attacked a much more difficult problem, a problem as much philosophical as mathematical. He died before he could complete the work. A generation of mathematicians has been pondering over how to follow up on his insights. As of this writing, this problem is still unsolved, and, as I shall show in the final chapters of this book, if it remains unsolved, the whole of the statistical approach to science may come crashing down from the weight of its own inconsistencies.

Kolmogorov's final problem was the question: What does probability mean in real life? He had produced a satisfactory mathematical theory of probability. This meant that the theorems and methods of probability were all internally self-consistent. The statistical model of science leaps out of the purely mathematical realm and applies these theorems to real-life problems. To do so, the abstract mathematical model that Kolmogorov proposed for probability theory has to be identified with some aspect of real life. There have been literally hundreds of attempts to do this, each one providing a different meaning to probability in real life and each of them subject to criticism. The problem is very important. The interpretation of the mathematical conclusions of statistical analysis depends upon how you identify these axioms with real-life situations.

In Kolomogorov's axiomization of probability theory, we assume there is an abstract space of elementary things called "events." Sets of events in this space can be measured the same way we measure the area of a porch deck or the volume of a refrigerator. If this measure on the abstract space of events fulfills certain axioms, then it is a probability space. To use probability theory in real life, we have to identify this space of events and do so with sufficient specificity to allow us to actually calculate probability measurements on that space. What is this space when an experimental scientist uses a statistical model to analyze the results? William Sealy Gosset proposed that the space was the set of all possible outcomes of the experiment, but he was unable to show how to calculate probabilities on it. Unless we can identify Kolmogorov's abstract space, the probability statements that emerge from statistical analyses will have many different and sometimes contrary meanings.

For instance, suppose we run a clinical trial to examine the efficacy of a new treatment for AIDS. Suppose the statistical analysis shows that the difference between the old and the new treatments is significant. Does this mean the medical community can be sure the new treatment will work on the next AIDS patient? Does this mean it will work for a certain percentage of AIDS patients? Does it merely mean that, only in the highly selected patient population of the study, there appears to be an advantage for the new treatment?

Finding the real-life meaning of probability has usually been approached by proposing real-life meaning for Kolmogorov's abstract probability space. Kolmogorov took another tack. Combining ideas from the second law of thermodynamics, early work by Karl Pearson, tentative attempts by several American mathematicians to find a mathematical theory of information, and work by Paul Lévy on the laws of large numbers, he produced a series of papers, starting in 1965, that wiped away the axioms and his solution to the mathematical problem and treated probability as . . .

On October 20, 1987, Andrei Nikolaevich Kolmogorov died, vibrant with life and pouring out original ideas to his last days—and no one has been able to pick up the threads that he left.

Afternote on the Failures of Soviet Statistics

Although Kolmogorov and his students made major contributions to the mathematical theories of probability and statistics, the Soviet Union gained little from the statistical revolution. Why this was the case provides an example of what happens when the government knows the "correct" answer to all questions.

During the final days of the czar and in the early years of the Russian Revolution, there was a considerable amount of statistical activity in Russia. Russian mathematicians were fully aware of the work being published in Britain and Europe. Papers by Russian mathematicians and agronomists appeared in *Biometrika*. The revolutionary government set up a Central Statistical Administration, and there were similar statistical administrations in the individual Soviet republics. The Central Statistical Administration published a journal of statistical activity, *Vestnik Statistiki*, which contained summaries of articles appearing in English- and German-language journals. As late as 1924, *Vestnik Statistiki* published a description of the application of statistical designs to agricultural research.

With the coming of the Stalin terror in the 1930s, the cold hand of orthodox Communist theory fell on these activities. The theoreticians of the party (the theologians of their religion according to Saints Marx and Lenin, to quote Chester Bliss; see chapter 8) looked upon statistics as a branch of social science. Under Communist doctrine, all social science was subordinate to central planning. The mathematical concept of a random variable lies at the heart of statistical methods. The Russian word for *random variable* translates as "accidental magnitude." To the central planners and theoreticians, this was an insult. All industrial and social activity in the Soviet Union was planned according to the theories of Marx and Lenin. Nothing could occur by accident. Accidental magnitudes might describe things observed in capitalist economies—not in Russia. The applications of mathematical statistics were quickly stifled. S. S. Zarkovic, in a historical review of Soviet statistics published in *The Annals of Mathematical Statistics* in 1956, described this with some delicacy:

> In subsequent years, political considerations became an increasingly pronounced factor in the development of Russia's statistics. This brought about the gradual disappearance of the use of theory in the practical activity of the statistical administration. In the late thirties, *Vestnik Statistiki* began to close its pages to papers in which statistical problems were dealt with mathematically. At the end of the thirties, they disappeared completely and have not appeared since. The result of this trend was that statisticians abandoned practice to continue their work at universities and other scientific institutions where they pursued statistics under the name of some other subjects. Officially, A. N. Kolmogorov, N. V. Smirnov, V. I. Romanovsky and many others are mathematicians divorced from statistics. A very interesting example is E. Slutsky, who enjoyed worldwide renown as one of the forerunners of econometrics. He gave up statistics to embark on a new career in astronomy. . . . According to official views, statistics became an instrument for planning the national economy. Consequently it represents a social science or, in other words, a class science. The law of large numbers, the idea of random deviations, and

> everything else belonging to the mathematical theory of statistics were swept away as the constituent elements of the false universal theory of statistical science.

Nor did the official views stop at statistics. Stalin embraced a quack biologist named Trofim D. Lysenko, who rejected the genetic theory of inheritance and claimed that plants and animals could be molded by their environment without having to inherit traits. Biologists who tried to follow R. A. Fisher's work in mathematical genetics were discouraged or even sent to prison. As orthodox theory descended on Soviet statistics, the numbers generated by the Central Statistical Administration and its successors became more and more suspect. Under central planning, the rich farmlands of Ukraine and Belorussia were becoming muddy wastes. Vast amounts of poorly built machines that would not work and consumer items that fell apart poured out of Russian factories. The Soviet Union was having trouble feeding its populace. The only economic activity that worked was the black market. The central government still turned out false, optimistic statistics in which the exact level of economic activity was hidden behind reports dealing with rates of change and rates of rates of change.

While American mathematicians like Norbert Wiener were using the theorems of Kolmogorov and Alexander Ya. Khintchine on stochastic processes to further the American war effort, while Walter Shewhart and others at the U.S. Bureau of Standards were showing American industry how to use statistical methods of quality control, while the agricultural output of American, European, and some Asian farms was increasing by leaps and bounds, Soviet factories continued turning out worthless machines and Soviet farms were unable to feed the nation.

Only in the 1950s, with the coming to power of Nikita Khrushchev, did this cold hand of official theory begin to lift, and tentative attempts were made to apply statistical methods to industry and agriculture. The official "statistics" continued to be filled with lies and elaborate obfuscations, and all efforts to publish journals that dealt with applied statistics resulted in a few irregularly printed issues. The extension of modern statistical models to Russian industry had to wait for the complete collapse of the Soviet Union and its system of central planning in the late 1990s.

Perhaps there is a lesson to be learned from this.

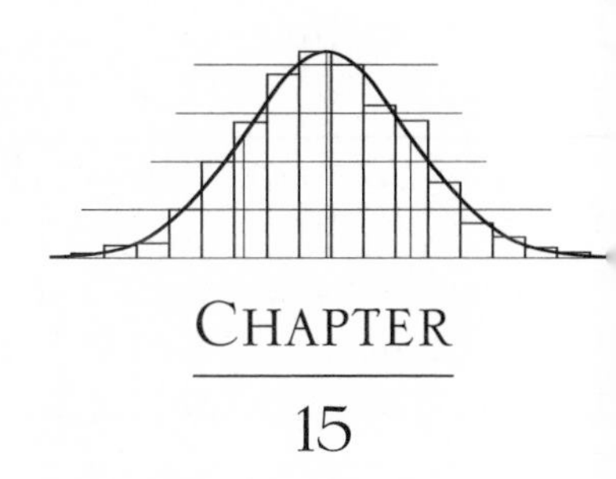

CHAPTER

15

THE WORM'S-EYE VIEW

Florence Nightingale, a legendary English Victorian figure, was a terror to the members of Parliament and the British army generals whom she confronted. There is a tendency to think of her simply as the founder of the nursing profession, a gentle, self-sacrificing giver of mercy, but the real Florence Nightingale was a woman with missions. She was also a self-educated statistician.

One of Nightingale's missions was to force the British army to maintain field hospitals and supply nursing and medical care to soldiers in the field. To support her position, she plowed through piles of data from the army files. Then she appeared before a royal commission with a remarkable series of graphs. In them, she showed how most of the deaths in the British army during the Crimean War were due to illnesses contracted outside the field of battle, or that occurred long after action as a result of wounds suffered in battle but left unattended. She invented the pie chart as a means of displaying her message.

When she tired of fighting the obtuse, seemingly ignorant army generals, she would retreat to the village of Ivington, where she would always be welcomed by her friends, the Davids. When the young David couple had a daughter, they named her Florence Nightingale David. Some of Florence Nightingale's feistiness and pioneering spirit seem to have transferred to her namesake. F. N. David (the name under which she published ten books and more than 100 papers in scientific journals) was born in 1909 and was five years old when World War I interrupted what would have been the normal course of her education. Because the family lived in a small country village, her first schooling consisted of private lessons with the local parson. The parson had some peculiar ideas of education for the young Florence Nightingale David. He noted that she had already learned some arithmetic, so he started her on algebra. He felt that she had already learned English, so he started her on Latin and Greek. At age ten, she transferred to formal schooling.

When it came time for Florence Nightingale David to go to college, her mother was appalled at her desire to attend University College, London. University College had been founded by Jeremy Bentham (whose mummified body sits in formal clothes in the cloisters of the college). The college was designed for "Turks, Infidels, and such as do not profess the Thirty-nine Articles." Until its founding, only those who professed the Thirty-nine Articles of faith of the established Church of England were allowed to teach or study at the universities in England. Even when David was preparing for college, University College still had the reputation of being a hotbed of dissenters. "My mother was having fits by this time about my going to London . . . disgrace and iniquity and that sort of thing." So she went to Bedford College for Women in London.

"I didn't like it very much," she said much later in a recorded conversation with Nan Laird of the Harvard School of Public Health. "But what I did like was I went to the theatre every night. If you were a student, you could go to the Old Vic for sixpence. . . .

The mummified body of Jeremy Bentham, preserved in the cloisters of University College, London

I had a great time." At school, she continued, "I did nothing but mathematics for three years, and I didn't like that very much. I didn't like the people and I suppose I was a rebel in those days. But I don't look back on it fondly."

What could she do with all that mathematics when she graduated? She wanted to become an actuary, but the actuarial firms would take only men. Someone suggested that she see a fellow named Karl Pearson at University College who, her informant had heard, had something to do with actuaries or something like that. She walked over to University College and "I crashed my way in to see Karl Pearson." He took a liking to her and got her a scholarship to continue her studies as his research student.

WORKING FOR K. P.

Working for Karl Pearson, F. N. David was set to computing the solutions of complicated difficult multiple integrals, calculating the distribution of the correlation coefficient. This work produced her first book, *Tables of the Correlation Coefficient*, which was finally published in 1938. She did all these calculations, and much more during those years, on a hand-cranked mechanical calculator known as a Brunsviga. "I estimated that I turned that hand Brunsviga roughly 2 million times. . . . Before I learned to manipulate long knitting needles [to unjam the machine] . . . I was always jamming the damn thing. When you jammed it, you were supposed to go tell the professor and then he would tell you what he thought of you; it was really rather awful. Many was the time I had jammed the machine and had gone home without telling him." Although she admired him and was to spend a great deal of time with him in his final years, David was terrified of Karl Pearson in the early 1930s.

She was also a daredevil and used to take a motorcycle on cross-country races.

> I had a hell of a crash one day into a 16-foot wall, which had glass on the top, and I pitched over and hurt my knee. I was in my office one day and I was miserable and [William S.] Gosset came and he said, "Well, you better take up flyfishing," because he was an ardent flyfisher. He invited me to his house. There was him [sic] and Mrs. Gosset and various children in the house in Henden. He taught me to throw a fly and he was very kind.

David was at University College, London, when Neyman and young Egon Pearson began looking at Fisher's likelihood function, angering old Karl Pearson, who thought it all nonsense. Egon was afraid of angering his father further, so rather than submit this first work to his father's journal, *Biometrika*, he and Neyman began a

Jerzy Neyman with two of his "ladies," Evelyn Fix (left) *and F. N. David* (right)

new journal, *Statistical Research Memoirs*, which ran for two years (and in which F. N. David published several papers). Then Karl retired and Egon took over the editorship of *Biometrika* and ended the *Memoirs*. F. N. David was there when the "old man" (as he was called) was being usurped by his son and R. A. Fisher. She was there when young Jerzy Neyman was just starting his research in statistics. "I think the period between the 1920s and 1940 was really seminal in statistics," she said. "And I saw all the protagonists from a worm's-eye point of view."

David called Karl Pearson a wonderful lecturer. "He lectured so well, you would sit there and let it all soak in." He was also tolerant of interruptions from students, even if one of them spotted a mistake, which he would quickly correct. Fisher's lectures, on the other hand, "were awful. I couldn't understand anything. I wanted to ask him a question, but if I asked him a question, he wouldn't answer it because I was a female." So she'd sit next to one of the male students from America and push his arm, saying, "Ask him! Ask him!" "After Fisher's lecture, I would go spend about three hours in the library trying to understand what he was up to."

When Karl Pearson retired in 1933, F. N. David went with him as his only research assistant. As she wrote:

> Karl Pearson was an extraordinary person. He was in his seventies and we would have worked all day on something and go out of the college at 6:00. On one occasion he was going home and I was going home and he said to me, "Oh, you just might have a look at the elliptic integral tonight. We shall want it tomorrow." And I hadn't the nerve to tell him that I was going off with a boyfriend to the Chelsea Arts Ball. So I went to the Arts ball and came home at four to five in the morning, had a bath, went to the University, and then had it ready when he came in at nine. One's silly when one's young.

A few months before Karl Pearson died, F. N. David came back to the biometrical laboratory and worked with Jerzy Neyman. Neyman was surprised she did not have her doctorate. At his urging, she took her last four published papers and submitted them as a dissertation. She was asked later if her status changed as a result of being granted a Ph.D. "No, no," she said. "I was just out the twenty-pound entrance fee."

Looking back at those days, she noted, "I'm inclined to think that I was brought in to keep Mr. Neyman quiet. But it was a tumultuous time because Fisher was upstairs raising hell and there was Neyman on one side and K. P. on the other and Gosset coming in every other week." Her reminiscences of those years are much too modest. She was far more than a worm "brought in to keep Mr. Neyman quiet." Her published papers (including one very important one cowritten by Neyman on a generalization of a seminal theorem by A. A. Markov, a Russian mathematician of the early twentieth century) advanced the practice and theory of statistics in many fields. I can pull books off my shelf from almost every branch of statistical theory and find references to papers by F. N. David in all of them.

War Work

When war broke out in 1939, David worked at the Ministry of Home Security, trying to anticipate the effects of bombs that might be dropped on population centers like London. Estimates of the number of casualties, the effects of bombs on electricity, water, and sewage systems, and other potential problems were determined from statistical models she built. As a result, the British were prepared for the German blitz of London in 1940 and 1941 and were able to maintain essential services while saving lives.

Toward the end of the war, as she wrote:

> I was flown over in one of the American bombers to Andrews Air Force Base. I came over to have a look at the first big digital computers that they built. . . . It was a Nissen hut [called a Quonset hut in the United States], about 100 yards long, and down the center were a whole lot of duckboards, on bits of wood you could run on. Either side, about every few feet, there were two winking monsters, and in the ceiling was [sic] nothing but fuses. Every thirty seconds or so, the GI would run down the duckboard with his face up to the sky and push in a fuse. . . . When I got back, I was telling somebody about it . . . and they said, "Well, you better sit down and learn the language." And I said, "Not on your nelly. If I do, that's all I will ever do for the rest of my life and no, I'm not going to—somebody else can!"

Egon Pearson was not the domineering person his father was, and he initiated a new policy that had the chairmanship of the biometrics department rotate among the other members of the faculty. About the time F. N. David rotated to the chairmanship, she had begun working on *Combinatorial Chance*, a book that is one of the classics in the literature. It is a remarkably clear exposition of complicated methods of counting, known as "combinatorics." When she was asked about this book, in which exceedingly complicated ideas are presented with a single underlying approach that makes them much easier to understand, she replied:

> All my life, I have had this beastly business that I start something and then I get bored with it. I had the idea of combinatorics and I had worked on them for a long time, long before I knew Barton [D. E. Barton, the coauthor of her book, later to become professor of computer science at University College] or taught Barton . . . but I took him on because it was time that the thing was finished. So we got to work and he did all the fancy work, proceeding to limits and things like that. He was a good chap. We wrote quite a lot of papers together.

She eventually came to the United States, where she was on the faculty at the University of California, Berkeley, and succeeded Neyman as chair of the department. She left Berkeley to found and chair the statistics department at the University of California, Riverside, in 1970. She "retired" in 1977 at age sixty-eight and became an active professor emeritus and research associate in biostatistics at Berkeley. The interview that provided many of the quotations in this chapter took place in 1988. She died in 1995.

In 1962, F. N. David published a book entitled *Games, Gods, and Gambling*. Here is her description of how it came about:

> I had lessons in Greek when I was young and . . . I got interested in archaeology when I had a colleague in archaeology who was busy digging in one of the deserts,

> I think. Anyway, he came to me and said, "I have walked about the desert and I have plotted where these shards were. Tell me where to dig for the kitchen midden." Archaeologists don't care about gold and silver, all they care about are the pots and pans. So I took his map and I thought about it and I thought this is exactly like the problem of the V-bombers. Here you have London and here you have bombs landing and you want to know where they come from so you can assume a normal bivariate surface and predict the major axes. That's what I did with the shard map. It's curious there's a sort of unity among problems, don't you think? There's only about a half dozen of them that are really different.

And Florence Nightingale David contributed to the literature of all of them.

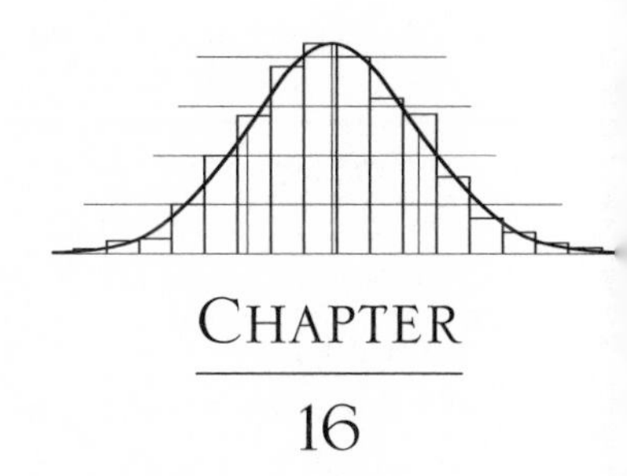

Chapter
16

Doing Away With Parameters

During the 1940s, Frank Wilcoxon, a chemist at American Cyanamid, was bothered by a statistical problem. He had been running hypothesis tests comparing the effects of different treatments, using "Student"'s[1] t-tests and Fisher's analyses of variance. This was the standard way of analyzing experimental data at that time. The statistical revolution had completely taken over the scientific laboratory, and books of tables for interpreting these hypothesis tests were on every scientist's shelves. But Wilcoxon was concerned about what often appeared to be a failure of these methods.

He might run a series of experiments where it was obvious to him that treatments differered in effect. Sometimes the t-tests would declare significance, and sometimes they would not. It often happens, when running an experiment in chemical engineering, that the chemical reactor, where the reaction takes place, is not sufficiently warmed up at the

[1]Recall that "Student" was the pseudonym of William S. Gosset, who developed the first small-sample statistical tests.

beginning of the sequence of experimental trials. It may happen that a particular enzyme begins to vary in its ability to react. The result is an experimental value that appears to be wrong. It is often a number that is much too large or much too small. Sometimes, it is possible to identify the cause of this outlying result. Sometimes, the result is an outlier, differing drastically from all the other results, but there is no obvious reason for this.

Wilcoxon looked at the formulas for calculating t-tests and analyses of variance and realized that these outliers, these extreme and unusual values, greatly influenced the results, causing "Student"'s t-test to be smaller than it would be otherwise. (In general, large values of the t-test lead to small p-values.) It was tempting to eliminate the outlier from the set of observations and calculate the t-test from the other values. This would introduce problems into the mathematical derivation of the hypothesis tests. How would the chemist know whether a number is really an outlier? How many outliers would have to be eliminated? Could the chemist continue to use the tables of probability for the standard test statistics if outliers have been eliminated?

Frank Wilcoxon went on a literature search. Surely the great mathematical masters who had produced the statistical methods had seen this problem before! He could find no reference to it. Wilcoxon thought he saw a way out of the problem. It involved tedious calculations based on combinations and permutations of the observed numbers (F. N. David's combinatorics are mentioned in the previous chapter). He began working out a method for calculating those combinatoric values.

Oh, but this was foolish! Why should a chemist like Wilcoxon have to work out these simple but tedious calculations? Surely, somebody in statistics had done this already! Back he went to the statistical literature to find this previous paper. He found no such paper. Mainly to check on his own mathematics, he submitted a paper to the journal *Biometrics* (not to be confused with Pearson's *Biometrika*). He still believed that this could not be original work and he depended upon the referees to know where this had been published before and reject his paper. By rejecting it, they would also notify him about these other references. However, as far as the referees and editors could determine, this was original work. No one had ever thought of this before, and his paper was published in 1945.

What neither Wilcoxon nor the editors of *Biometrics* knew was that an economist named Henry B. Mann and a graduate student in statistics at Ohio State University named D. Ransom Whitney were working on a related problem. They were trying to order statistical distributions so that one might say, in some sense, that the distribution of wages in the year 1940 was less than the distribution of wages in 1944. They came up with a method of ordering that involved a sequence of simple but tedious counting methods.

This led Mann and Whitney to a test statistic whose distribution could be computed from combinatoric arithmetic—the same type of computation as Wilcoxon's. They published a paper describing their new technique in 1947, two years after Wilcoxon's paper had appeared. It was quickly shown that the Wilcoxon test and the Mann-Whitney test were closely related and produced the same p-values. Both these test statistics involved something new. Until the Wilcoxon paper, it was thought that all test statistics would have to be based on estimates of parameters of distributions. This was a test, however, that did not estimate any parameter. It compared the observed scatter of data with what might have been expected from purely random scatter. It was a nonparametric test.[2]

In this way, the statistical revolution moved a step beyond Pearson's original ideas. It could now deal with distributions of measurements without making use of parameters. Unknown to many in the West, in the late 1930s, Andrei Kolmogorov in the Soviet

---

[2]Actually, as further justification of Stigler's law of misonomy, Wilcoxon was not the first to suggest nonparametric methods. Work by Karl Pearson in 1914 seems to suggest some of these ideas. However, the nonparametric approach was not fully understood to be such a drastic revolution until Wilcoxon's work in this field.

Union and a student of his, N. V. Smirnov, had investigated a different approach to the comparison of distributions that did not make use of parameters. The work by Wilcoxon and Mann and Whitney had opened a new window of mathematical investigation by directing attention to the underlying nature of ordered ranks, and the Smirnov-Kolmogorov work was soon included.

FURTHER DEVELOPMENTS

Once a new window has been opened in mathematical research, investigators begin to look through it in different ways. This original work by Wilcoxon was soon followed by alternative approaches. Herman Chernoff and I. Richard Savage discovered that the Wilcoxon test could be looked upon in terms of the expected average values of ordered statistics; and they were able to expand the nonparametric test into a set of tests involving different underlying distributions, none of them requiring the estimation of a parameter. By the early 1960s, this class of tests (now referred to as "distribution-free tests") was the hot topic in research. Doctoral students filled in little corners of the theory to provide their theses. Meetings were devoted exclusively to this new theory. Wilcoxon continued to work in the area, expanding on the range of tests as he developed extremely clever algorithms for the combinatoric calculations.

In 1971, Jaroslav Hájek from Czechoslovakia produced a definitive textbook that provided a unifying view for the entire field. Hájek, who died in 1974 at age forty-eight, discovered an underlying generalization for all nonparametric tests and linked this general approach to the Lindeberg-Lévy conditions of the central limit theorem. This is frequently the way of mathematical research. In some sense, all mathematics are linked, but the exact nature of those links and the insights to exploit them often take many years to emerge.

As he pursued the implications of his statistical discovery, Frank Wilcoxon left his original field of chemistry and managed the statistical services group at American Cyanamid and at its Lederle Labs division. In 1960, he joined the faculty of the Department of Statistics at Florida State University, where he proved himself to be a well-regarded teacher and researcher and oversaw several doctoral candidates. When he died in 1965, he left behind a legacy of students and statistical innovation that continue to have a remarkable effect on the field.

Unsolved Problems

The development of nonparametric procedures may have led to a burst of activity in this new field. However, there was no obvious link between the parametric methods that were used before this and the nonparametric methods. There were two unsolved questions:

1. If the data have a known parametric distribution, like the normal distribution, how badly will the analysis go wrong if we use nonparametric methods?

2. If the data do not quite fit a parametric model, how far off from that model must the data be before the nonparametric methods are the better ones to use?

In 1948, the editors of the *Annals of Mathematical Statistics* received a paper from an unknown mathematics professor at the University of Tasmania, on the island off the southern coast of Australia. This remarkable paper solved both problems. Edwin James George Pitman had published three earlier papers in the *Journal of the Royal Statistical Society* and one in the *Proceedings of the Cambridge Philosophical Society* that, in retrospect, laid the foundations of his later work but that had been ignored or forgotten. Except for those four papers, Pitman, who was fifty-two years old when he submitted his article to the *Annals*, was unpublished and unknown.

E. J. G. Pitman had been born in Melbourne, Australia, in 1897. He attended the University of Melbourne as an undergraduate, but his schooling was interrupted by the First World War, when he served two years in the army. He returned to complete his degree. "In those days," he wrote later, "there were no graduate schools in mathematics in Australian universities." Some of the universities provided scholarship money for their better students to go on to graduate study in England, but the University of Melbourne did not. "When I left the University of Melbourne after four years of study there, I had had no training in research; but I thought that I had learned to study and use mathematics, and that I would be willing to tackle any problem that arose. . . ." The first problem was how to earn a living.

The University of Tasmania was looking for someone to teach mathematics. Pitman applied and was named professor of mathematics. The entire department consisted of the new professor and a part-time lecturer. The department was required to give courses in undergraduate mathematics to students in all other departments, and the new professor was busy with a course load that took almost all his time. When the board of trustees decided to hire a full-time mathematics professor, one of the board members had heard that there was a new branch of mathematics called statistics; so the new applicant was asked if he would be prepared to teach a course in statistics (whatever that was).

Pitman replied, "I cannot claim to have any special knowledge of the Theory of Statistics; but, if appointed, I would be prepared to lecture on this subject in 1927." He had no special knowledge, nor any other type of knowledge, of the theory of statistics. At Melbourne, he had taken a course in advanced logic, during which the professor devoted a couple of lectures to statistics. As Pitman put it, "I decided then and there that statistics was the sort of thing that I was not interested in, and would never have to bother about."

Young E. J. G. Pitman arrived in the fall of 1926 in Hobart, Tasmania, with only an undergraduate degree to support his title of professor, at a small provincial school that was about as far away as one could get from the intellectual ferment in London and Cambridge. "I did not publish anything," he wrote, "until 1936. There were two main reasons for the delay in publication; the work load I had to carry, and the nature of my upbringing," by which he meant his lack of training in methods of mathematical research.

By 1948, when he sent his remarkable paper to the *Annals of Mathematical Statistics*, the mathematics department at the University of Tasmania had grown. It now had one professor (Pitman), an associate professor, two lecturers, and two tutors. They taught a wide range of mathematics, both applied and theoretical. Pitman gave twelve lectures a week and covered classes on Saturdays. He now had some support for his research. Starting in 1936, the Commonwealth government began to provide 30,000 pounds a year for the promotion of scientific research in the universities of Australia. This was allocated to different states on a population basis; because Tasmania was one of the smaller states, this amounted to 2,400 pounds a year for the entire university. How much of this Pitman got, he did not say.

Gradually, Pitman engaged in different types of research. His first published paper dealt with a problem in hydrodynamics. His next three papers investigated highly specific aspects of hypothesis testing theory. These papers were not remarkable in themselves, but they were Pitman's learning theses. He was exploring how to develop ideas and how to relate mathematical structures to each other.

By the time he began to work on his 1948 paper, Pitman had developed a clear line of reasoning about the nature of statistical hypothesis tests and the interrelationships between the older (parametric) and the newer (nonparametric) tests. With his new methods, he attacked the two outstanding problems.

What he found surprised everyone. Even when the original assumptions are true, the nonparametric tests were almost as good as the parametric tests. Pitman was able to answer the first question: How badly do we do if we use nonparametric tests in a situation where we know the parametric model and should be using a specific parametric test? Not badly at all, said Pitman.

The answer to the second question was even more surprising. If the data do not fit the parametric model, how far off from the parametric model must they be for the nonparametric tests to be better? Pitman's calculations showed that with only slight deviations from the parametric model, the nonparametric tests were vastly better than the parametric ones.

It looked as if Frank Wilcoxon, the chemist who was sure somebody had made this simple discovery before him, had stumbled onto a veritable philosopher's stone. Pitman's results suggested that all hypothesis tests should be nonparametric ones. Pearson's discovery of the statistical distributions based on parameters was only a first step. Now statisticians were able to deal with statistical distributions without worrying about specific parameters.

There are subtleties within subtleties in mathematics. Deep within their seemingly simple approaches, Wilcoxon, Mann and Whitney, and Pitman had made assumptions about the distributions of the data. It would take another twenty-five years for these assumptions to be understood. The first disturbing problem was discovered by R. R. Bahadur and L. J. ("Jimmie") Savage at the University of Chicago in 1956. When I showed the Bahadur and Savage paper to a friend of mine from India a few years ago, he remarked at the congruence of their names. *Bahadur* means "warrior" in Hindi. It took a warrior and a savage to strike the first blow to the theory of nonparametric statistical tests.

The problems that Savage and Bahadur uncovered came from the very problem that had first suggested nonparametric tests to Wilcoxon: the problem of outliers. If the outliers are rare and completely "wrong" observations, then nonparametric methods reduce their influence on the analysis. If the outliers are part of a systematic contamination of the data, then shifting to nonparametric methods may only make matters worse. We shall investigate the problem of contaminated distributions in chapter 23.

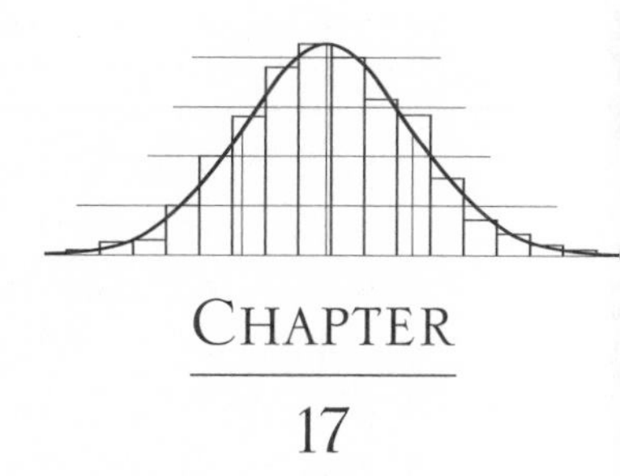

CHAPTER

17

WHEN PART IS BETTER THAN THE WHOLE

To Karl Pearson, probability distributions could be examined by collecting data. He thought that if he collected enough data, he could expect them to be representative of all such data. The correspondents for *Biometrika* would select hundreds of skulls from ancient graveyards, pour shot into them to measure the cranial capacity, and send Pearson these hundreds of numbers. A correspondent would travel to the jungles of Central America and measure the lengths of arm bones on hundreds of natives, sending these measurements to Pearson's biometrical laboratory.

There was a basic flaw in Pearson's methods, however. He was collecting what is now called an "opportunity sample." The data were those that were most easily available. They did not have to be truly representative of the entire distribution. The graves that were opened to find the cranial capacity of skulls were those that happened to have been found. The unfound ones could differ in some unknown way.

A specific example of this failure of opportunity sampling was discovered in India in the early 1930s. Bales of jute were assembled on the dock in Bombay for shipment to Europe. To determine the value of the jute, each bale was sampled and the quality of the jute determined from the sample. The sampling was done by plunging a hollow circular blade into the bale and pulling out a small amount in the core of the blade. In the packing and shipment of the bales, the outside of the bales tended to deteriorate and the inner parts tended to become more and more compacted, often including some frozen parts in the wintertime. The sampler would push the hollow knife into the bale, but it would be deflected from the denser part of the bale, and the sample would tend to consist almost entirely of the outer and damaged region. The opportunity sample was biased in favor of finding inferior jute when the quality of the bale was much higher.

Professor Prasanta Chandra Mahalanobis, head of the Department of Physics in the Presidency College, Calcutta, often used this example (which he had discovered when working for the railroad that shipped the jute to the dock) to show why opportunity samples were not to be trusted. Mahalanobis came from a wealthy family of Calcutta merchants and could afford a period of graduate and postgraduate study as he pursued his interests in science and mathematics. During the 1920s, he traveled to England and studied under both Pearson and Fisher. Students like F. N. David had to subsist on scholarship aid, but Mahalanobis lived the life of the grand seigneur while he pursued his studies. He returned to head the Department of Physics at Presidency College. Soon afterward, in 1931, he used his own funds to set up the Indian Statistical Institute on the grounds of one of his family estates.

At the Indian Statistical Institute, he trained a group of brilliant Indian mathematicians and statisticians, many of whom went on to make important contributions to the field—people like S. N. Roy, C. R. Rao, R. C. Bose, P. K. Sen, and Madan Puri, among others. One of Mahalanobis's interests lay in this question of how to produce an appropriately representative sample of data. It was clear that, in many situations, it was almost impossible to get all the measurements in a set. For instance, the population of India is so great that, for years, no attempt was made to get a complete census on a single day—as is tried in the United States. Instead, the complete Indian census takes over a year, as different regions of the country are counted in different months. Because of this, the Indian census can never be accurate. There are births and deaths, migrations, and changes of status that occur during the time of the census. No one will ever know exactly how many people there are in India on a given day.[1]

Mahalanobis reasoned that it might be possible to estimate the characteristics of the larger population if one could gather a small sample that was adequately representative of the larger. At this point, we run into two possible approaches. One is to construct what is known as a "judgment sample." In a judgment sample, whatever is known about the population is used to select a small group of individuals who are chosen to represent different groups in the larger population. The Nielsen ratings to determine how many people are watching TV shows are created from a judgment sample. Nielsen Media Research selects families based on their socioeconomic status and the region of the country in which they live.

A judgment sample seems, at first glance, to be a good way to get a representative sample of the larger population. But it has two major faults. The first is that the sample is representative only if we are absolutely sure that we know enough about the larger population to be able to find specific subclasses that can be represented. If we knew that much about the larger population, we would

[1]In the United States, an attempt is made to count all the people on a given day for the decennial census. However, investigations of the 1970 census and those that followed showed that the complete count tended to miss many people and double-count others. Furthermore, the missing persons are usually from specific socioeconomic groups, so it cannot be assumed that they are "similar" to citizens who were counted. It might also be said, even of the United States, that no one will ever know exactly how many people there are on a given day.

probably not have to sample, since the questions we ask of the sample are those that are needed to divide the larger population into homogeneous groups. The second problem is more troublesome. If the results of the judgment sample are wrong, we have no way of knowing how far from the truth they are. In the summer of 2000, Nielsen Media Research was criticized for not having enough Hispanic families in their sample and underestimating the number of families watching Spanish-language TV.

Mahalanobis's answer was the random sample. We use a randomizing mechanism to pick individuals out of the larger population. The numbers we get from this random sample are most likely wrong, but we can use the theorems of mathematical statistics to determine how to sample and measure in an optimum way, making sure that, in the long run, our numbers will be closer to the truth than any others. Furthermore, we know the mathematical form of the probability distribution of random samples, and we can calculate confidence bounds on the true values of the things we wish to estimate.

Thus, the random sample is better than either the opportunity sample or the judgment sample, not because it guarantees correct answers but because we can calculate a range of answers that will contain the correct answer with high probability.

THE NEW DEAL AND SAMPLING

The mathematics of sampling theory developed rapidly during the 1930s, some of it at the Indian Statistical Institute under Mahalanobis, some of it from two papers by Neyman in the late 1930s, and some of it by a group of eager young university graduates who gathered in Washington, D.C., during the early days of the New Deal. Many of the practical problems of how to sample from a large population of people were addressed and solved by these young New Dealers at the Commerce and Labor Departments of the federal government.

A young man or woman who received a bachelor's degree in the years 1932 to 1939 often walked out of the university into a world where there were no jobs. The Great Depression saw to that. Margaret Martin, who grew up in Yonkers, New York, and went to Barnard College, and who eventually became an official of the U.S. Bureau of the Budget, wrote:

> When I graduated in June of 1933 I couldn't find any job. . . . A friend of mine who graduated one year later in 1934 felt very fortunate. She'd gotten a job selling at the B. Altman department store; she worked 48 hours a week and earned $15. But, even those jobs were relatively scarce. We had an occupation officer, Miss Florence Doty, at Barnard, and I went to her to talk about the possibilities of going to Katherine Gibbs, a secretarial school. I didn't know where I'd find the money, but I thought that would be a skill that could at least earn something. Miss Doty . . . was not an easy person to get along with, and a lot of the students stood very much in awe of her. . . . She just turned on me, "I would never recommend that you take a secretarial course! If you learn to use a typewriter, and show that you can use a typewriter, you will never do anything else but use a typewriter. . . . You should be looking for a professional position."

Martin eventually found her first job in Albany, as a junior economist in the office of research and statistics of the New York State Division of Placement and Unemployment, and used that job as a springboard to graduate studies.

Other newly graduated young people went directly to Washington. Morris Hansen went to the Census Bureau in 1933, with an undergraduate degree in economics from the University of Wyoming. He used his undergraduate mathematics and a hasty reading of Neyman's papers to design the first comprehensive survey of unemployment. Nathan Mantel took his new City College of New York (CCNY) degree in biology and went to the National Cancer Institute. Jerome Cornfield, a history major at CCNY, took a job as an analyst at the Department of Labor.

It was an exciting period to be in government. The nation lay prostrate, with much normal economic activity idle, and the new government in Washington was looking for ideas on how to get things started again. First they had to know how bad things were throughout the country. Surveys of employment and economic activity were begun. For the first time in the nation's history, an attempt was being made to determine exactly what was happening in the country. It was an obvious place for sample surveys.

These eager young workers initially had to overcome the objections of those who did not understand the mathematics. When one of the earlier surveys at the Department of Labor showed that less than 10 percent of the population received almost 40 percent of the income, it was denounced by the U.S. Chamber of Commerce. How could this be true? The survey had contacted less than one-half of 1 percent of the working population, and these people were chosen by random means! The Chamber of Commerce had its own surveys, taken from the opinions of its own members about what was happening. This new survey was dismissed by the chamber as inaccurate, since it was only a random collection of data.

In 1937, the government attempted to get a full count of the unemployment rate, and Congress authorized the Unemployment Census of 1937. The bill, as passed by Congress, called for everyone who was unemployed to fill out a registration card and deliver it to the local post office. At that time, estimates of the number of unemployed ranged from three to fifteen million, and the only good counts were a few random surveys that had been taken in New York. A group of young sociologists, led by Cal Dedrick and Fred Stephan in the Census Bureau, realized that there would be many unemployed who would not respond and the result would yield numbers filled with unknown errors. It was decided that the first serious random survey across the entire country should be run. With young Morris Hansen designing the survey, the bureau chose 2 percent of all the postal routes at random. The postal carriers on those routes handed out questionnaires to every family on their routes.

Even with a 2 percent sample, the Census Bureau was overwhelmed with the huge number of questionnaires. The U.S. Postal Service attempted to organize them and make initial tabulations. The questionnaire had been designed to pick up detailed information about the demographics and the work history of the respondents, and no one knew how to examine such large amounts of detailed information. Recall that this was before the computer, and the only aids to pencil-and-paper tabulations were hand-operated mechanical calculators. Hansen contacted Jerzy Neyman, whose papers had formed the basis of the survey's design. In Hansen's words, Neyman pointed out that "we didn't have to know and match all the cases and understand all the relationships" to find answers to the most important questions. Using Neyman's advice, Hansen and his coworkers set aside most of the complicated and confusing details of the questionnaires and counted the numbers of unemployed.

It took a series of careful studies in the Census Bureau, under Hansen, to prove that these small random surveys were much more accurate than the judgment samples that had been used before. Eventually, the U.S. Bureau of Labor Statistics and the Census Bureau led the way into a new world of random sampling. George Gallup and Louis Bean took these methods into the region of political polling.[2] For the 1940 census, the Census Bureau initiated elaborate plans for sample surveys within the overall census. There was a young, newly hired statistician at the bureau named William Hurwitz. Hansen and Hurwitz became close collaborators and

[2]In the late 1960s, I attended a session where Louis Bean was a speaker. He described those early years, when he and Gallup began using surveys to advise political candidates. Gallup went public with a syndicated column, the Gallup Poll. Bean continued to do private polling, but he warned Gallup that he might set up his own column and call it the Galloping Bean Poll.

friends; they issued a series of important and influential papers, culminating in the 1953 textbook *Sample Survey Methods and Theory* (written with a third author, William Madow). The Hansen and Hurwitz papers and text became so important to the field of sample surveys, and were quoted so often, that many workers in the field came to believe that there was one person named Hansen Hurwitz.

JEROME CORNFIELD

Many of the new young workers who arrived in Washington during the New Deal went on to become major figures in government and in academia. Some of them were too busy creating new mathematics and statistical methods to go for graduate degrees. A prime example is Jerome Cornfield. Cornfield participated in some of these early surveys at the Bureau of Labor Statistics and then moved to the National Institutes of Health. He published papers jointly with some of the leading figures in academia. He solved the mathematical problems involved in case-control studies. His scientific papers range from work on random sample theory to the economics of employment patterns, the investigation of tumors in chickens, problems in photosynthesis, and effects of environmental toxins on human health. He created many of the statistical methods that have now become standard in the fields of medicine, toxicology, pharmacology, and economics.

One of Cornfield's most important achievements was in the design and initial analysis of the Framingham Study, begun in 1948. The idea was to take Framingham, Massachusetts, as a "typical town," measure a large number of health variables on everyone in the town, and then follow these people for a number of years. The study has now been running for over fifty years. It has had a "Perils of Pauline" existence as, from time to time, attempts have been made to cut its funding in the interests of budget reduction in government. It remains a major source of information on the long-term effects of diet and lifestyles on heart disease and cancer.

To analyze the first five years of data from the Framingham Study, Cornfield ran into fundamental problems that had not been addressed in the theoretical literature. Working with faculty members at Princeton University, he solved these problems. Others went on to produce papers on the theoretical development he started, but Cornfield was satisfied to have found a method. In 1967, he was a coauthor of the first medical article emerging from the study, the first article to show the effects of elevated cholesterol on the probability of heart disease.

I was on a committee with Jerry Cornfield, convened in 1973 as part of a set of hearings before a Congressional committee. During a break in our work, Cornfield was called to the phone. It was Wassily Leontief, an economist at Columbia University, calling to say that he had just been awarded the Nobel Prize in economics and wanted to thank Cornfield for the role Jerry had played in their work, which led to this prize. This work had originated in the late 1940s when Leontief had come to the Bureau of Labor Statistics for help.

Leontief believed that the economy could be broken down into sectors, like farming, steel manufacturing, retailing, and so forth. Each sector uses material and services from the other sectors to produce material or a service, which it supplies to those other sectors. This interrelationship can be described in the form of a mathematical matrix. It is often called an "input–output analysis." When he first began investigating this model at the end of World War II, Leontief went to the Bureau of Labor Statistics to help gather the data he needed. To assist him, the bureau assigned a young analyst who was working there at the time, Jerome Cornfield.

Leontief could break the economy down into a few broad sectors, such as putting all manufacturing in one sector, or he could subdivide the sectors into more specific ones. The mathematical theory of input–output analysis requires that the matrix that describes the economy have a unique inverse. That meant that the matrix, once assembled, had to be subjected to a mathematical procedure called "inverting the matrix." At that time, before the widespread availability of computers, inverting a matrix was a difficult and tedious procedure on a calculator. When I was in graduate school, each of us had to invert a matrix—I suspect as a kind of rite of passage "for the good of our souls." I remember trying to invert a 5 × 5 matrix and taking several days, most of which I spent locating my mistakes and redoing what I had done wrong.

Leontief's initial set of sectors led to a 12 × 12 matrix, and Jerry Cornfield proceeded to invert that 12 × 12 matrix to see if there was a unique solution. It took him about a week, and the end result was the conclusion that the number of sectors had to be expanded. So, with trepidation, Cornfield and Leontief began subdividing the sectors until they ended with the simplest matrix they thought would be feasible, a 24 × 24 matrix. They both knew this was beyond the capacity of a single human being. Cornfield estimated that it would take him several hundred years of seven-day workweeks to invert a 24 × 24 matrix.

During World War II, Harvard University had developed one of the first, very primitive computers. It used mechanical relay switches and would often jam. There was no longer any war work for it, and Harvard was looking for applications for its monstrous machine. Cornfield and Leontief decided to send their 24 × 24 matrix to Harvard where its Mark I computer would go through the tedious calculations and compute the inverse. When they sought to pay for this project, the process was stopped by the accounting office of the Bureau of Labor Statistics. The government had a policy at that time; it would pay for goods but not for services. The theory was that the government had all kinds of experts working for it. If something had to be done, there should be someone in government who could do it.

They explained to the government accountant that, while this was theoretically something that a person could do, no one would be able to live long enough to do it. The accountant was sympathetic, but he could not see a way around the regulation. Cornfield then made a suggestion. As a result, the bureau issued a purchase order for capital goods. What capital goods? The invoice called for the bureau to purchase from Harvard "one matrix, inverted."

Economic Indices

The work of these young men and women who rushed into government during the early days of the New Deal continues to be of fundamental importance to the nation. This work led to the regular series of economic indicators that are now used to fine-tune the economy. These indicators include the Consumer Price Index (for inflation), the Current Population Survey (for unemployment rates), the Census of Manufacturing, the intermediate adjustments of Census Bureau estimates of the nation's population between decennial censuses, and many other less well-known surveys that have been copied and are used by every industrial nation in the world.

In India, P. C. Mahalanobis became a personal friend of Prime Minister Jawaharlal Nehru in the early days of the new government of India. Under his influence, Nehru's attempts to imitate the central planning of the Soviet Union were often modified by carefully conducted sample surveys, which showed what really was happening to the new nation's economy. In Russia, the bureaucrats produced false figures of production and economic activity to flatter the rulers, which encouraged the more foolish excesses of their central economic plans. In India, good estimates of the truth were always available. Nehru and his successors may not have liked it, but they had to deal with it.

In 1962, R. A. Fisher went to India. He had been there many times before at Mahalanobis's invitation. This was a special occasion. There was a great meeting of the world's leading statisticians to commemorate the thirtieth anniversary of the founding of the Indian Statistical Institute. Fisher, Neyman, Egon Pearson, Hansen, Cornfield, and others from the United States and Europe were there. The sessions were lively, for the field of mathematical statistics was still in ferment, with many unsolved problems. The methods of statistical analysis were penetrating to every field of science. New techniques of analysis were constantly being proposed and examined. There were four scientific societies devoted to the subject and at least eight major journals (one of which had been founded by Mahalanobis).

When the conference closed, the attendees went their separate ways. As they arrived home, they heard the news. R. A. Fisher had died of a heart attack on the boat returning him to Australia. He was seventy-two years old. His collected scientific papers fill five volumes, and his seven books continue to influence all that is being done in statistics. His brilliant original accomplishments had come to an end.

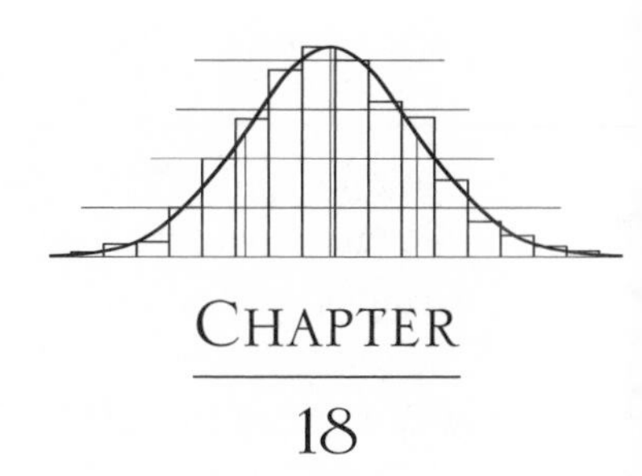

CHAPTER

18

DOES SMOKING CAUSE CANCER?

In 1958, R. A. Fisher published a paper entitled "Cigarettes, Cancer, and Statistics" in the *Centennial Review*, and two papers in *Nature* entitled "Lung Cancer and Cigarettes?" and "Cancer and Smoking." He then pulled these together, along with an extensive preface, in a pamphlet entitled "Smoking: the Cancer Controversy. Some Attempts to Assess the Evidence." In these papers, Fisher (who was often photographed smoking a pipe) insisted that the evidence purported to show that smoking caused lung cancer was badly flawed.

Nor was Fisher alone in his criticisms of the smoking/cancer studies at that time. Joseph Berkson, the head statistician at the Mayo Clinic and a leader among American biostatisticians, questioned the results. Jerzy Neyman had raised objections to the reasoning used in the studies that associated lung cancer and cigarette smoking. Fisher was the most strident in his criticism. As the evidence accumulated over the next few years, and both Berkson and Neyman appeared satisfied that the relationship was proved, Fisher remained adamant, actually accusing some of the leading researchers of doctoring their data. It became an embarrassment to many statisticians. At that time, the cigarette companies were denying the validity of the studies, pointing out that they were only "statistical correlations" and that there was no proof that cigarettes caused lung cancer. On the surface, it appeared that Fisher was agreeing with them. His arguments had the air of a polemic. Here, for instance, is a paragraph from one of his papers:

> The need for such scrutiny [of the research that appeared to show the relationship] was brought home to me very forcibly about a year ago in an annotation published by the British Medical Association's *Journal*, leading up to the almost shrill conclusion that it was necessary that every device of modern publicity should be employed to bring home to the world at large this terrible danger. When I read that, I wasn't sure that I liked "all the devices of modern publicity," and it seemed to me that a moral distinction ought to be drawn at this point. . . . [It] is not quite so much the work of a good citizen to plant fear in the minds of perhaps a hundred million smokers throughout the world—to plant it with the aid of all the means of modern publicity backed by public money—without knowing for certain that they have anything to be afraid of in the particular habit against which the propaganda is to be directed. . . .

Unfortunately, in his anger against the use of government propaganda to spread this fear, Fisher did not state his objections very clearly. It became the conventional wisdom that he was playing the role of a crotchety old man who did not want to relinquish his beloved pipe. In 1959, Jerome Cornfield joined with five leading cancer experts from the National Cancer Institute (NCI), the American Cancer Society, and the Sloan-Kettering Institute, to write a thirty-page paper that reviewed all the studies that had been published. They examined Fisher's, Berkson's, and Neyman's objections, along with objections raised by the Tobacco Institute (on behalf of the tobacco companies). They provided a carefully reasoned account of the controversy and showed how the evidence was overwhelmingly in favor of showing that "smoking is a causative factor in the rapidly increasing incidence of human epidermoid carcinoma of the lung."

That settled the issue for the entire medical community. The Tobacco Institute continued to pay for full-page advertisements in popular magazines, which questioned the association as being only a statistical correlation, but no articles that questioned this finding appeared after 1960 in any reputable scientific journal. Within four years, Fisher was dead. He could not continue the argument, and no one else took it up.

Is There Such a Thing as Cause and Effect?

Was it all a lot of nonsense put forward by an old man who wanted to smoke his pipe in peace, or was there something to Fisher's objections? I have read Fisher's smoking and cancer papers, and I have compared them to previous papers he had written on the nature of inductive reasoning and the relationship between statistical models and scientific conclusions. A consistent line of reasoning emerges. Fisher was dealing with a deep philosophical problem—a problem that the English philosopher Bertrand Russell had addressed in the early 1930s, a problem that gnaws at the heart of scientific thought, a problem that most people do not even recognize as a problem: What is meant by "cause and effect"? Answers to that question are far from simple.

Bertrand Russell may be remembered by many readers as a white-haired, grandfatherly looking but world-renowned philosopher, who lent his voice to the criticism of United States involvement in the war in Vietnam in the 1960s. By that time, Lord Russell had received both official and scholarly recognition as one of the great minds of twentieth-century philosophy. His first major work, written with Alfred North Whitehead—who was many years his senior—dealt with the philosophical foundations of arithmetic and mathematics. Entitled *Principia Mathematica*, it tried to establish the basic ideas of mathematics, like numbers and addition, on simple axioms dealing with set theory.

One of the essential tools of the Russell-Whitehead work was symbolic logic, a method of inquiry that was one of the great new creations of the early twentieth century. The reader may recall having studied Aristotelian logic with examples like "All men are mortal. Socrates is a man. Therefore, Socrates is mortal."

Although it has been studied for about 2,500 years, Aristotle's codification of logic is a relatively useless tool. It belabors the obvious, sets up arbitrary rules as to what is logical and what is not, and fails to mimic the use of logic in mathematical reasoning, the one place where logic has been used to produce new knowledge. While students were dutifully memorizing categorizations of logic based on Socrates's mortality and the blackness of raven feathers, the mathematicians were discovering new areas of thought, like calculus, with the use of logical methods that did not fit neatly into Aristotle's categories.

This all changed with the development of set theory and symbolic logic in the final years of the nineteenth century and early years of the twentieth. In its earliest form, the one that Russell and Whitehead exploited, symbolic logic starts with atoms of thought known as "propositions." Each proposition has a truth value called "T" or "F."[1] The propositions are combined and compared with symbols for "and," for "or," for "not," and for "equals." Because each of the atomic propositions has a truth value, any combination of

[1]Note the abstract nature of this. "T," of course, means "True," and "F" means "False." By using apparently meaningless symbols, the mathematicians are able to think of variations on the ideas. Suppose, for instance, we proposed three truth values: "T," "F," and "M" (for "Maybe"). What does this do to the mathematics? The use of purely abstract symbols has led to fascinating complexities in symbolic logic, and the subject has remained an active area of mathematical research for the past ninety years.

them has a truth value, which can be computed via a series of algebraic steps. On this simple foundation, Russell, Whitehead, and others were able to build combinations of symbols that described numbers and arithmetic and seemed to describe all types of reasoning.

All except one! There seemed to be no way to create a set of symbols that meant "A causes B." The concept of cause and effect eluded the best efforts of the logicians to squeeze it into the rules of symbolic logic. Of course, we all know what "cause and effect" means. If I drop a glass tumbler on the bathroom floor, this act causes it to break. If the master restrains the dog whenever it goes in the wrong direction, this act causes the dog to learn to go in the right direction. If the farmer uses fertilizer on his crops, this act causes the crops to grow bigger. If a woman takes thalidomide during the first trimester of her pregnancy, this act causes her child to be born with attenuated limbs. If another woman suffers pelvic inflammation, it was because of the IUD she used.[2] If there are very few women in senior management positions at the ABC firm, it was caused by prejudice on the part of the managers. If my cousin has a hair-trigger temper, this was caused by the fact that he was born under the sign of Leo.

As Bertrand Russell showed very effectively in the early 1930s, the common notion of cause and effect is an inconsistent one. Different examples of cause and effect cannot be reconciled to be based on the same steps of reasoning. There is, in fact, no such

[2]In the case of *Marder* v. *G. D. Searle*, which ran through the federal courts in the 1980s, the plaintiff claimed her illness resulted from her IUD. As proof, the plaintiff presented epidemiological evidence showing an apparent increase in frequency of pelvic inflammation among women who used IUDs. The defendant presented a statistical analysis that computed 95 percent confidence bounds on the relative risk (the probability of getting the disease with the IUD divided by the probability of getting the disease without the IUD). The confidence bounds ran from .6 to 3.0. The jury was deadlocked. The judge ruled in favor of the defendant, stating: "[It] is particularly important to be assured that an inference of causation is based upon at least a reasonable probability of causation." There is an unstated assumption here that probability can be defined as personal probability. Although the opinion tried to distinguish between "causation" and "statistical correlation," the confusion this involves, and that also occurs in the rulings of higher courts, points up the basic inconsistency involved in the concept of cause and effect, which Russell discussed fifty years before.

thing as cause and effect. It is a popular chimera, a vague notion that will not withstand the batterings of pure reason. It contains an inconsistent set of contradictory ideas and is of little or no value in scientific discourse.

MATERIAL IMPLICATION

In place of cause and effect, Russell proposed the use of a well-defined concept from symbolic logic, called "material implication." Using the primitive notions of atomic propositions and the connecting symbols for "and," "or," "not," and "equals," we can produce the concept that proposition A implies proposition B. This is equivalent to the proposition that not B implies not A. This begins to sound a little like the paradox that lies behind Bayes's theorem (which we looked at in chapter 13). But there are very deep differences, which we will examine in a later chapter.

In the late nineteenth century, the German physician Robert Koch proposed a set of postulates needed to prove that a certain infective agent caused a specific disease. These postulates required:

1. Whenever the agent could be cultured, the disease was there.
2. Whenever the disease was not there, the agent could not be cultured.
3. When the agent was removed, the disease went away.

With some redundancy, Koch was stating the conditions for material implication. This may be adequate to determine that a particular species of bacteria caused the infectious disease. When it comes to something like smoking and cancer, however, Koch's postulates are of little value. Let us consider how well the connection between lung cancer and cigarette smoking fits Koch's postulates (and hence Russell's material implication). The agent is a history of cigarette smoking. The disease is human epidermoid carcinoma of the lung. There are cigarette smokers who do not get lung cancer. Koch's first postulate is not met. There are some people who get lung cancer who claim that they have not been smokers. If we are to believe their claims, Koch's second postulate is not met. If we restrict the type of cancer to small oat-cell carcinoma, the number of nonsmokers with this disease appears to be zero, so maybe the second postulate has been met. If we take the agent away, that is, if the patient stops smoking, the disease may still come, and Koch's third postulate is not met.

If we apply Koch's postulates (and with them Russell's material implication), then the only diseases that meet them are acute conditions that have been caused by specific infective agents that can be cultured from the blood or other fluids of the body. This does not hold for heart disease, diabetes, asthma, arthritis, or cancer in other forms.

Cornfield's Solution

Let us return to the 1959 paper by Cornfield and five prominent cancer specialists.[3] One by one, they describe all the studies that had been run on the subject. First there was a study by Richard Doll and A. Bradford Hill,[4] published in the *British Medical Journal* in 1952. Doll and Hill had been alarmed by the rapid rise in the number of patients dying from lung cancer in the United Kingdom. They located several hundred cases of such patients and matched them to similar patients (same age, sex,

[3]The coauthors were William Haenszel of the National Cancer Institute (NCI), E. Cutler Hammond of the American Cancer Society, Abraham Lilienfeld of the School of Hygiene and Public Health, Johns Hopkins University, Michael Shimkin of the NCI, and Ernst Wynder of the Sloan-Kettering Institute. However, the paper was proposed and organized by Cornfield. In particular, Cornfield wrote the sections that carefully examine and refute Fisher's arguments.

[4]In spite of the fact that R. A. Fisher chose to savage the work of Hill and Doll in particular, both men were prominent in the extension of Fisher's methods to medical research. Almost single-handedly, Hill convinced the British medical community that useful information can be gained only from studies that follow the principles of Fisher's experimental design. The name of Richard Doll, later to become Regius Professor of Medicine at Oxford University, is synonymous with the conversion of modern clinical research to statistical models.

socioeconomic status) who had been admitted to the same hospitals at the same time, but who did not have lung cancer. There were almost ten times as many smokers among the lung cancer patients than there were among the others (called the "controls" in such a study). By the end of 1958, there were five other studies of this nature, using patients in Scandinavia, the United States, Canada, France, and Japan. All of them showed the same results: a much greater percentage of smokers among the cancer patients than among the controls.

These are called "retrospective studies." They start with the disease and work backward to see what prior conditions are associated with the disease. They need controls (patients without the disease) to be sure that these prior conditions are associated with the disease and not with some more general characteristic of the patients. These controls can be criticized as not matching the disease cases. One prominent retrospective study was run in Canada on the effects of artificial sweeteners as a cause of bladder cancer. The study seemed to show an association between artificial sweeteners and bladder cancer, but a careful analysis of the data showed that the disease cases were almost all from low socioeconomic classes and the controls were almost all from upper socioeconomic classes. This meant that the disease cases and the controls were not comparable. In the early 1990s, Alvan Feinstein and Ralph Horvitz at the Yale Medical School proposed very rigid rules for running such studies to ensure that the cases and controls match. If we apply the Feinstein-Horvitz rules to these retrospective case-control cancer and smoking studies, all of them fail.

An alternative approach is the prospective one. In such a study, a group of individuals is identified in advance. Their smoking histories are carefully recorded, and they are followed to see what will become of them. By 1958, three independent prospective studies had been run. The first (reported by the same Hill and Doll who had done the first retrospective study) involved 50,000 medical doctors in the United Kingdom. Actually, in the Hill and Doll study, the subjects were not followed over a long period of time. Instead, the 50,000 doctors were interviewed about their health habits, including their smoking habits, and followed for five years, as many of them began to come down with lung cancer. Now the evidence did more than suggest a relationship. They were able to divide the doctors into groups depending upon how much they smoked. The doctors who smoked more had greater probabilities of having lung cancer. This was a dose response, the key proof of an effect in pharmacology. In the United States, Hammond and Horn ran a prospective study (published in 1958) on 187,783 men, whom they followed for four months. They also found a dose response.

There are some problems with prospective studies, however. If the study is small, it may be dealing with a particular population. It may not make sense to extrapolate the results to a larger population. For instance, most of these early prospective studies were done with males. At that time, the incidence of lung cancer in females was too low to allow for analysis. A second problem with prospective studies is that it may take a long time for enough of the events (lung cancer) to occur to allow for sensible analysis. Both these problems are dealt with by following a large number of people. The large number gives credence to the suggestion that the results hold for a large population. If the probability of the event is small in a short period of time, following a large number of people for a short period of time will still produce enough events to allow for analysis.

The second Doll and Hill study used medical doctors because it was believed that their recollection of smoking habits could be relied upon and because their belonging to the medical profession made it virtually certain that all the lung cancers that occurred in the group would be recorded. Can we extrapolate the results from educated, professional doctors to what would happen to a dock-hand with less than a high school education? Hammond and Horn used almost 200,000 men in hopes that their sample would be more representative—at the risk of getting less than accurate information. At this point, the reader may recall the objection to Karl Pearson's samples of data because they were opportunity samples. Weren't these also opportunity samples?

To answer this objection, in 1958, H. F. Dorn studied the death certificates from three major cities and followed up with interviews of the surviving families. This was a study of all deaths, so it could not be considered an opportunity sample. Again, the relationship between smoking and lung cancer was overwhelming. However, the argument could be made that the interviews with surviving family members were flawed. By the time this study was run, the relationship between lung cancer and smoking was widely known. It was possible that surviving relatives of patients who had died of lung cancer would be more likely to remember that the patient had been a smoker than would relatives of patients who had died from other diseases.

Thus it is with most epidemiological studies. Each study is flawed in some way. For each study, a critic can dream up possibilities that might lead to bias in the conclusions. Cornfield and his coauthors assembled thirty epidemiological studies run before 1958 in different countries and concentrating on different populations. As they point out, it is the overwhelming consistency across these many studies, studies of all kinds, that lends credence to the final conclusion. One by one, they discuss each of the objections. They consider Berkson's objections and show how one study or another can be used to address them. Neyman suggested that the initial retrospective studies could be biased if the patients who smoked lived longer than the nonsmokers and if lung cancer was a disease of old age. Cornfield et al. produced data about the patients in the studies to show that this was not a sensible description of those patients.

They addressed in two ways the question of whether the opportunity samples were nonrepresentative. They showed the range of patient populations involved, increasing the likelihood that the conclusions held across populations. They also pointed out that, if the cause and effect relationship holds as a result of fundamental biology, then the patients' different socioeconomic and racial backgrounds would be irrelevant. They reviewed toxicology studies, which showed carcinogenic effects of tobacco smoke on lab animals and tissue cultures.

This paper by Cornfield et al. is a classic example of how cause is proved in epidemiological studies. Although each study is flawed, the evidence keeps mounting, as one study after another reinforces the same conclusions.

Smoking and Cancer Versus Agent Orange

A contrast to this can be seen in the attempts to indict Agent Orange as a cause for health problems Vietnam War veterans have suffered in later life. The putative agents of cause are contaminants in the herbicide that was used. Almost all studies have dealt with the same small number of men exposed in different ways to the herbicide. Studies in other populations did not support these findings. In the 1970s, an accident in a chemical factory in northern Italy resulted in a large number of people being exposed to much higher levels of the contaminant, with no long-term effects. Studies of workers on New Zealand turf farms who were exposed to the herbicide suggested an increase in a specific type of birth defect, but the workers were mostly Maoris, who have a genetically related tendency toward that particular birth defect.

Another difference between the smoking and the Agent Orange studies is that the putative consequences of smoking are highly specific (epidermoid carcinoma of the lung). The events that were supposedly caused by Agent Orange exposure consisted of a wide range of neurological and reproductive problems. This runs contrary to the usual finding in toxicology that specific agents cause specific types of lesions. For the Agent Orange studies, there is no indication of a dose response, but there are insufficient data to determine the different doses to which individuals have been exposed. The result is a muddied picture, where objections like those of Berkson, Neyman, and Fisher have to go unaddressed.

With the analysis of epidemiological studies, we have moved a long way from the highly specific exactitude of Bertrand Russell and material implication. Cause and effect are now imputed from many flawed investigations of human populations. The relationships are statistical, where changes in the parameters of distributions appear to be related to specific causes. Reasonable observers are expected to integrate a large number of flawed studies and see the underlying common threads.

PUBLICATION BIAS

What if the studies have been selected? What if all that is available to the observer is a carefully selected subset of the studies that were actually run? What if, for every positive study that is published, a negative study was suppressed? After all, not every study gets published. Some never get written up because the investigators are unable or unwilling to complete the work. Some are rejected by journal editors because they do not meet the standards of that journal. All too often, especially when there is some controversy associated with the subject, editors are tempted to publish that which is acceptable to the scientific community and reject that which is not acceptable.

This was one of Fisher's accusations. He claimed that the initial work by Hill and Doll had been censored. He tried for years to have the authors release detailed data to back up their conclusions. They had only published summaries, but Fisher suggested that these summaries had hidden inconsistencies that were actually in the data. He pointed out that, in the first Hill and Doll study, the authors had asked if the patients who smoked inhaled when they smoked. When the data are organized in terms of "inhalers" and "noninhalers," the noninhalers are the ones with an excess of lung cancer. The inhalers appear to have less lung cancer. Hill and Doll claimed that this was probably due to a failure on the part of the respondents to understand the question. Fisher scoffed at this and asked why they didn't publicize the real conclusions of their study: that smoking was bad for you, but if you had to smoke, it was better to inhale than not to inhale.

To Fisher's disgust, Hill and Doll left that question out of their investigation when they ran their prospective study on medical doctors. What else was being carefully selected? Fisher wanted to know. He was appalled that the power and money of government was going to be used to throw fear into the populace. He considered this no different from the use of propaganda by the Nazis to drive public opinion.

Fisher's Solution

Fisher had also been influenced by Bertrand Russell's discussion of cause and effect. He recognized that material implication was inadequate to describe most scientific conclusions. He wrote extensively on the nature of inductive reasoning and proposed that it was possible to conclude something in general about life on the basis of specific investigations, provided that the principles of good experimental design were followed. He showed that the method of experimentation, where treatments were randomly assigned to subjects, provided a logically and mathematically solid basis for inductive inference.

The epidemiologists were using the tools Fisher developed for the analysis of designed experiments, such as his methods of estimation and tests of significance. They were applying these tools to opportunity samples, where the assignment of treatment was not from some random mechanism external to the study but an intricate part of the study itself. Suppose, he mused, that there was something genetic that caused some people to be smokers and others not to smoke. Suppose, further, that this same genetic disposition involved the occurrence of lung cancer. It was well known that many cancers have a familial component. Suppose, he said, this relationship between smoking and lung cancer was because each was due to the same event, the same genetic disposition. To prove his case, he assembled data on identical twins and showed that there was a strong familial tendency for both twins to be either smokers or nonsmokers. He challenged the others to show that lung cancer was not similarly genetically influenced.

On one side there was R. A. Fisher, the irascible genius who put the whole theory of statistical distributions on a firm mathematical setting, fighting one final battle. On the other side was Jerry Cornfield, the man whose only formal education was a bachelor's degree in history, who had learned his statistics on his own, who was too busy creating important new statistics to pursue a higher degree. You cannot prove anything without a randomized experimental design, said Fisher. Some things do not lend themselves to such designs, but the accumulation of evidence should prove the case, said Cornfield. Both men are now deceased, but their intellectual descendants are still with us. These arguments resound in the courts, where attempts are made to prove discrimination on the basis of outcomes. They play a role in attempts to identify the harmful results of human activity on the biosphere. They are there whenever great issues of life and death arise in medicine. Cause and effect are not so simple to prove, after all.

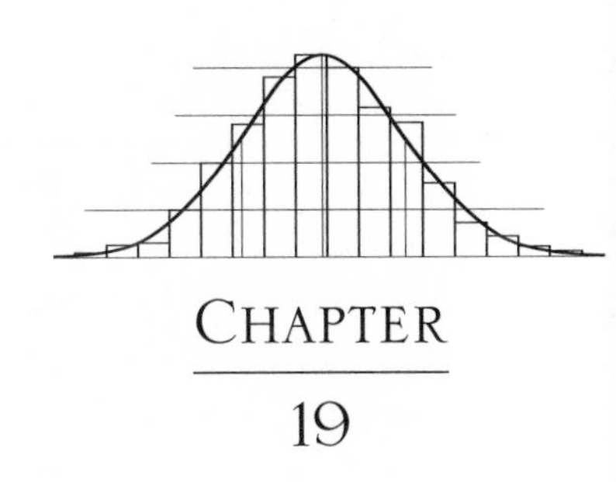

CHAPTER

19

IF YOU WANT THE BEST PERSON . . .

Late in the summer of 1926, George W. Snedecor left the University of Kentucky, where he had just earned a doctorate in mathematics, packed his few belongings in a suitcase, and set out by auto for the University of Iowa, where he had heard there was an opening for an assistant professor of mathematics. Unfortunately, he knew nothing about the geography of Iowa and eventually found himself in Ames, Iowa, the home of Iowa State University, rather than in Iowa City, where he would have found the University of Iowa. No, he was told, they did not have an opening for a mathematics professor here at Ames. This was an agricultural college, and whatever mathematics education was needed was taken care of by a small staff already on hand. However, their agronomists had been reading some papers by this fellow, R. A. Fisher, in England, and they would like to adopt his new methods. There was a lot of mathematics in those papers, and they could use somebody to help them understand the papers.

Snedecor, whose own education in mathematics had contained no courses in probability theory, stayed on in Ames to study these new developments and to found a statistical laboratory. Eventually, he created a department of statistics that was the first such academic department in the United States. He studied Fisher's papers and went on to review the work of Pearson, "Student," Edgeworth, Yates, von Mises, and others. While Snedecor did not contribute much in the way of original research, he was a great synthesizer. In the 1930s, he produced a textbook, *Statistical Methods*, first in mimeographed form and finally published in 1940, which became the preeminent text in the field. He improved on Fisher's *Statistical Methods for Research Workers* by including the basic mathematical derivations, by pulling similar ideas together, and by including an extensive set of tables that made it possible to calculate p-values and confidence intervals with a minimum of effort. In the 1970s, a review of citations in published scientific articles from all areas of science showed that Snedecor's *Statistical Methods* was the most frequently cited book.

Snedecor was also an effective administrator. He invited the leading figures in statistical research to spend summers at Ames. For most of the 1930s, Fisher himself visited for several weeks at a time, lecturing and consulting. The Statistical Laboratory and Department of Statistics at Ames, Iowa, became one of the most important centers of statistical research in the world. The list of men and women who were visiting professors at Ames during the years before World War II was a roster of the most distinguished people in the field.

Gertrude Cox came to study at Iowa State during this time. She had dreams of becoming a missionary and saving souls in far-off lands. For almost seven years after high school, she had dedicated her life to social service in the Methodist Church. She needed a university education to apply to the missionary service that interested her. Snedecor convinced her that statistics was more interesting, and she stayed on after she graduated to work with him at the

Statistical Laboratory. In 1931, she received the first master's degree in statistics granted by Iowa State, and Snedecor hired her to teach in his department. She became particularly interested in Fisher's theories of experimental design and taught the first courses on experimental design at Ames. Snedecor found a place for her in the graduate program in psychology at the University of California, where she continued her studies for two more years. She returned to Ames with her doctorate, and Snedecor put her in charge of the Statistical Laboratory.

In the meantime, the stream of prominent statisticians continued to flow through Ames, Iowa. William Cochran stopped for a while to take an appointment on the faculty. He joined Gertrude Cox in teaching experimental design courses (there were now several such courses), and together in 1950 they wrote a textbook on the subject, entitled *Experimental Designs*. Like Snedecor's *Statistical Methods*, Cochran and Cox's *Experimental Designs* takes the reader through the methods with a firm foundation of the mathematics behind it. It has a set of very useful tables to allow the experimenter to modify a design for specific situations and to analyze the results. The *Science Citation Index* publishes lists of citations from scientific journals each year. The *Index* is printed in small print with the citations ranged in five columns; Cochran and Cox's book usually takes up at least one full column each year.

THE CONTRIBUTIONS OF WOMEN

The reader will probably have noticed that, with the exception of Florence Nightingale David, all the statisticians described so far in this book have been male. The early years of statistical development were dominated by men. Many women were working in the field, but they were almost all employed in doing the detailed calculations needed for statistical analysis, and were indeed called "computers." Since extensive computations had to be made, all of them on hand-cranked calculators, this tedious work was usually delegated to women. Women tended to be more docile and patient, so went the belief, and could be depended upon more than men to check and recheck the accuracy of their calculations. A typical picture of the Galton Biometrical Laboratory under Karl Pearson would have Pearson and several men walking around, looking at output from the computers or discussing deep mathematical ideas, while all about them rows of women were computing.

As the twentieth century progressed, the situation began to change. Jerzy Neyman, in particular, was helpful and encouraging to many women, supervising their Ph.D. theses, publishing papers jointly with them, and finding prominent places for them in the academic community. By the 1990s, when I attended national meetings of the statistical societies, about half the attendees were women. Women are prominent in the American Statistical Association, the Biometric Society, the Royal Statistical Society, and the Institute of Mathematical Statistics. They are not yet equally represented with men, however. About 30 percent of the articles published in the statistical journals have one or more female authors, and only 13 percent of the members of the American Statistical Association who have been honored as fellows are women. This disparity is changing. In the last years of the twentieth century, the female half of the human race showed that it is capable of great mathematical activity.

This was not the case in 1940, when George Snedecor happened to meet Frank Graham, the president of the University of North Carolina, on a train. They sat together and had much to talk about. Graham had heard something of the statistical revolution, and Snedecor was able to fill him in on the great advances that had been made in agricultural and chemical research with statistical models. Graham was surprised to learn that the only full-fledged statistics department in the United States was at Iowa State. At Princeton University, Sam Wilks was developing a group of mathematical statisticians, but this was within the mathematics department. A similar situation existed at the University of Michigan with

Gertrude Cox, 1900–1978

Henry Carver.[1] Graham gave a great deal of thought to what he learned on that train ride.

Weeks later, Graham contacted Snedecor. He had convinced his sister school, North Carolina State University at Raleigh, that the time was ripe to set up a statistical laboratory and eventually a statistics department, similar to the ones at Ames. Could Snedecor recommend a man to head such a department? Snedecor sat down and drew up a list of ten men he thought would work out. He called in Gertrude Cox to verify his list and ask what she thought of it. She looked it over and asked, "What about me?"

[1]Henry Carver (1890–1977) was a lonely pioneer in the development of mathematical statistics as a respectable academic subject. From 1921 to 1941, he was the thesis adviser of ten doctoral candidates at the University of Michigan, all of whom were given topics in mathematical statistics. In 1930, he founded the journal *Annals of Mathematical Statistics*, and in 1938 he helped found the Institute of Mathematical Statistics, a scholarly organization that sponsored the *Annals*. The development of the *Annals* into a highly regarded journal is described in chapter 20.

Snedecor added a line to his letter: "These are the ten best men I can think of. But, if you want the best person, I would recommend Gertrude Cox."

Gertrude Cox turned out to be not only an excellent experimental scientist and a wonderful teacher, but also a remarkable administrator. She built a faculty of renowned statisticians who were also good teachers. Her students went on to take major positions in industry, academia, and government. She was treated by all of them with great respect and affection. When I first met her at a meeting of the American Statistical Association, I found myself sitting opposite a small, quiet elderly lady. When she talked, her eyes sparkled with enthusiasm as she would warm up to the matter under discussion, whether it was theoretical or involved a particular application. Her comments were salted with a delightful, if subdued, wit. I did not realize that she was suffering from leukemia, which would end her life soon after. In the years since her death, her former students gather each summer at the traditional joint meeting of the statistical societies where they sponsor a road race in her honor and raise money for scholarships in her name.

By 1946, Cox's Department of Applied Statistics had been so successful that Frank Graham was able to establish a department of mathematical statistics at the University of North Carolina in Chapel Hill, and soon afterward a department of biostatistics. The "triangle" of North Carolina State, University of North Carolina, and Duke University has become a center of statistical research, spawning private research firms that draw upon the expertise of these schools. The world that Gertrude Cox wrought has dwarfed the creation of her teacher, George Snedecor.

THE DEVELOPMENT OF ECONOMIC INDICATORS

Women have played an important role in the statistical components of government in America, serving in many senior positions at the Census Bureau, the Bureau of Labor Statistics, the National

Janet Norwood (right) *as Commissioner of Labor Statistics*

Center for Health Statistics, and the Bureau of Management and Budget. One of the most senior was Janet Norwood, who retired as Commissioner of the Bureau of Labor Statistics in 1991.

Norwood was attending Douglass College, the women's branch of Rutgers University in New Brunswick, New Jersey, when the United States entered World War II. Her boyfriend, Bernard Norwood, was leaving for the war, so they decided to get married. She was nineteen, he was twenty. He did not go overseas immediately, and they were able to be together. This marriage posed a problem for the sheltered world of Douglass College, however. They had never had married students before. Would the rules about male visitors hold for Janet and her husband? Would she need permission from her parents to leave campus and go to New York City to see him? This experience of being the first was to follow Janet Norwood. In 1949, she received a Ph.D. from Tufts University—the youngest person at that time ever to have done so. "From time to time," she wrote, "I have been the first woman elected to high office in organizations in which I have been active." She was the first woman to be named commissioner of labor statistics, a position she held from 1979 to 1991.

The administration may not have realized quite who they were putting in this position in 1979. Before Norwood became commissioner, it had been the practice of the Department of Labor that a representative of the policy arm of the department sat in on all reviews of the press releases planned by the Bureau of Labor Statistics. Norwood informed the representative that he would no longer be welcome at these meetings. She believed that the economic information produced by the bureau not only had to *be* accurate and nonpartisan, it had to *appear* to be so. She wanted all the bureau's activities completely insulated from the slightest possibility of political influence.

> I found that it was important to make clear that you were willing to resign on principle if the issue were important enough. . . . In government, you have the independence that you assert and insist upon. . . . Independence in government is not easy to accomplish. For example, how do you handle situations in which you have to correct the President of the United States? We did that.

Both Janet Norwood and her husband received Ph.D.s in economics. For the first few years of their married life, especially when her husband was involved in setting up the institutions of the European Common Market, she did not work outside the home, but raised two sons and wrote articles to keep active in her field. With the family settled in Washington, D.C., and with their second son in grade school, Janet Norwood looked for a position that would enable her to have some or all afternoons free, to be there when her son came home after school. Such a position opened at the Bureau of Labor Statistics, where she could arrange her work hours to be free three afternoons a week.

The Bureau of Labor Statistics would appear to be a minor bureau in the Department of Labor, which is itself a branch of government that seldom makes the headlines. What goes on in this lowly bureau, compared to the excitement of the White House or the State Department? It is, however, an essential cog in the machinery of government. Government has to run on information. The bright young men and women who came to Washington with the New Deal soon discovered that policy could not be made without some essential information about the state of the nation's economy, and such essential information was unavailable. An important innovation of the New Deal consisted of setting up the machinery necessary to produce this information.

The Bureau of Labor Statistics either conducts the surveys needed to produce this information or analyzes data accumulated from other departments, like the Census Bureau. Janet Norwood joined Labor Statistics in 1963. By 1970, she had risen to be in charge of the Consumer Price Index (CPI). The CPI is used to index social security payments, track inflation, and adjust most transfers of payment from the federal to the state governments. In 1978, the Bureau of Labor Statistics engaged in a major overhaul of the CPI, which Janet Norwood planned and supervised.

The CPI and other series generated by the bureau, over which Janet Norwood was to be the commissioner, involve complicated mathematical models, with relatively arcane parameters that make sense in econometric models but are often hard to explain to someone who lacks training in the mathematical side of economics.

The CPI is often quoted in newspapers as noting, for example, that inflation rose 0.2 percent last month. It is a complicated set of numbers, indicating the changes in price patterns across different sectors of the economy and in different regions of the country. It starts with the concept of a market basket. This is a set of goods and services that a typical family might purchase. Before the market basket can be assembled, sample surveys are taken to determine what families purchase and how often. Mathematical weights are computed to take into account that a family might purchase bread every week but a car only once every few years and a house less frequently.

Once the market basket and its associated mathematical weights have been assembled, employees of the bureau are sent out to randomly chosen stores, where they record the current prices for items on their list. The prices recorded are combined according to the mathematical weighting formulas and an overall number is computed that is, in some sense, the average cost of living for a family of a given size for that month.

In concept, the idea of an index to describe the average pattern of some economic activity is easy to understand. Attempting to *construct* such an index is more difficult. How does one take into account the appearance of a new product (like a home computer) on the market? How does one account for the possibility that the consumer might choose another but similar product if the price is too high (like choosing yogurt instead of sour cream)? The CPI and other measures of the nation's economic health are constantly being subjected to reexamination. Janet Norwood oversaw the last major overhaul of the CPI, but there will be others in the future.

The CPI is not the only index of the nation's economic health. There are other indices generated to cover manufacturing activity, inventories, and employment patterns. There are also social indicators, estimates of prison populations—all parameters associated with other noneconomic activity. These are, indeed, parameters in Karl Pearson's sense. They are parts of probability distributions, mathematical models, where the parameters do not describe a specific observable event but are "things" that govern the pattern of observable events. In this way, there is no family in the United States whose monthly costs are exactly equal to the CPI; and the unemployment rate cannot describe the real number of unemployed workers, which changes by the hour. Who, for that matter, is "unemployed"? Is it someone who has never been employed and is not now seeking work? Is it someone who is moving from one job to another with five weeks off and severance pay? Is it someone who is looking for only a few hours a week of work? The world of economic models is filled with arbitrary answers to such questions and involves a large number of parameters that can never be observed exactly but that interact with one another.

There is no R. A. Fisher to establish optimum criteria in the derivation of economic and social indicators. In each case, we are trying to reduce a complex interaction among people into a small collection of numbers. Arbitrary decisions have to be made. In the first census of unemployment in the United States, only heads of households (usually male) were counted. Current counts of the unemployed include anyone who was looking for work within the previous month. Janet Norwood, in overseeing a major overhaul of the CPI, had to reconcile the different opinions about similarly arbitrary definitions, and there will always be sincere critics who object to some of those definitions.

Women in Theoretical Statistics

Gertrude Cox and Janet Norwood, the two women described in this chapter, were primarily administrators and teachers. Women also played an important role in the development of theoretical statistics in the later half of the twentieth century. Recall L. H. C. Tippett's first asymptote of the extreme that is used to predict "hundred-year floods," as described in chapter 6. A version of that distribution, known as a "Weibull distribution," has found important uses in the aerospace industry. One problem with the Weibull distribution is that it does not meet Fisher's regularity conditions, and there are no clearly optimal ways of estimating its parameters. That is, there were no optimal ways until Nancy Mann at North American Rockwell discovered a link between the Weibull and a much simpler distribution, and developed the methods now used in that field.

Grace Wahba at the University of Wisconsin took a set of ad hoc methods of curve fitting, called "spline fits," and found a theoretical formulation that now dominates the statistical analysis of splines.

Yvonne Bishop was a member of a committee of statisticians and medical scientists who in the late 1960s were trying to determine if the widely used anesthetic halothane was the cause of an increase in liver failure among patients. The analysis was bedeviled by the fact that most of the data were in the form of counts of events. During the previous ten years, attempts had been made to organize complicated multidimensional tables of counts like those in this halothane study, but none had been particularly successful. Previous workers had suggested approaching such tables in a way that was similar to Fisher's analysis of variance, but that work was incomplete. Bishop took it up and examined the theoretical ramifications, establishing criteria for estimation and interpretation. Having polished the technique in the halothane study, she published a definitive text on it. "Log-linear models," as this method has come to be called, are now the standard first step in most sociological studies.

Since the days of Snedecor and Cox, the "best person" has frequently been a woman.

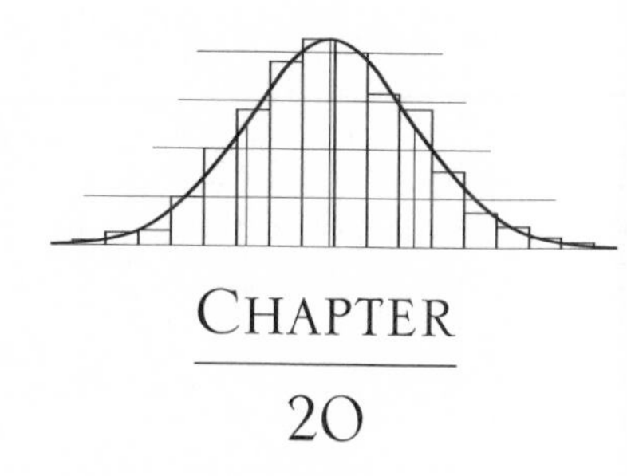

CHAPTER

20

JUST A PLAIN TEXAS FARM BOY

In the late 1920s, when Samuel S. Wilks left the family farm in Texas to study at the University of Iowa, mathematical research was scaling the heights of beautiful abstraction. Purely abstract fields like symbolic logic, set theory, point set topology, and the theory of transfinite numbers were sweeping through the universities. The level of abstraction was so great that any inspiration from real-life problems that might have given rise to the initial ideas in these fields had long since been lost. Mathematicians plunged into the axioms that the ancient Greek Euclid had declared were the foundations of mathematics; and they found unstated assumptions behind those axioms. They scrubbed mathematics clean of such assumptions, explored the fundamental building blocks of logical thought, and emerged with remarkable, seemingly self-contradictory ideas, like space-filling curves and three-dimensional shapes that touched everywhere and nowhere at the same time. They investigated the different orders of infinity and "spaces" with fractional dimensions. Mathematics was on the upsurge of an all-encompassing wave of sheer abstract thought, completely divorced from any sense of reality.

Nowhere was this drive into abstraction beyond practicality as great as in the mathematics departments of American universities. The publications of the American Mathematical Society were being recognized among the top echelon of mathematical journals in the world, and American mathematicians were pushing at the frontiers of abstractions beyond abstractions. As Sam Wilks was to state ruefully some years later, these departments, with their siren calls offering opportunities for pure thought, were siphoning off the best brains among American graduate students.

Sam Wilks's first graduate mathematics course at Iowa was taught by R. I. Moore, the most renowned member of the university's mathematics faculty. Moore's course in point set topology introduced Wilks to this wonderful world of impractical abstraction. Moore made it clear that he disdained useful work, and insisted that applied mathematics was on a level with washing dishes and sweeping streets. This is an attitude that has plagued mathematics since the time of the ancient Greeks. There is a story told of Euclid tutoring the son of a nobleman and going over a particularly beautiful proof of a theorem. In spite of Euclid's enthusiasm, the student seemed unimpressed and asked what use this might have. At which Euclid called over his slave and said, "Give the lad a copper. It seems he must gain from his knowledge."

Sam Wilks's bent for practical applications was answered by his thesis adviser, Everett F. Linquist, when he began looking for a Ph.D. thesis topic at Iowa. Linquist, who had worked in the mathematics of insurance, was interested in the newly developing field of mathematical statistics and proposed a problem from that area to Wilks. At that time, there was something slightly disreputable about mathematical statistics, at least among the mathematics departments of American and European universities. R. A. Fisher's great pioneering work had been published in "out-of-the-way" journals like the *Philosophical Transactions of the Royal Society of Edinburgh*. *The Journal of the Royal Statistical Society* and *Biometrika* were looked down upon as publications in which statistical collections of numbers were tabulated. Henry Carver at the University of Michigan had started a new journal entitled the *Annals of Mathematical Statistics*, but its standards were much too low for most mathematicians to take notice of it. Linquist suggested an interesting problem in abstract mathematics that emerged from a method of measurement used in educational psychology. Wilks solved this problem and used it for a doctoral thesis, and the results were published in the *Journal of Educational Psychology*.

To the world of pure mathematics, this was not much of an achievement. The field of educational psychology was well below their horizon of interest. But a Ph.D. thesis is supposed to be a first tentative step into the world of research, and few students are expected to make important contributions with their theses. Wilks went to Columbia University for a year of postgraduate study (in the course of which he was expected to increase his ability to handle the cold, rarefied abstractions of important mathematics). In the fall of 1933, he arrived at Princeton University, having taken a position as an instructor in mathematics.

STATISTICS AT PRINCETON

The Princeton mathematics department was immersed in the cold and beautiful abstractions as much as any other department in the United States. In 1939, The Institute of Advanced Studies was to be established nearby, and among its first members was Joseph H. M. Wedderburn, who had developed the complete generalization of all finite mathematical groups. Also at the institute were Hermann Weyl, who was famous for his work in nondimensional vector spaces, and Kurt Gödel, who had developed an algebra of metamathematics. These men influenced the Princeton faculty, which had its share of world-renowned mathematicians, prominent among them Solomon Lefshetz, who had opened the doors to the new abstract field of algebraic topology.[1]

In spite of the general bent toward abstraction among the Princeton faculty, Sam Wilks was fortunate to have Luther Eisenhart as chairman of the mathematics department. Eisenhart was interested in all kinds of mathematical endeavor and liked to encourage junior faculty members to follow their own inclinations. Eisenhart hired Wilks because he thought this new field of mathematical statistics held remarkable promise. Sam Wilks arrived in Princeton with his wife, pursuing a vision of applied mathematics that set him apart from the rest of the department. Wilks was a gentle fighter. He would disarm everyone with his "just folks" Texas farm boy attitude. He was interested in people as individuals and could persuade others to follow his vision, and he was extremely good at organizing work activities to accomplish difficult goals.

Wilks could often get to the heart of a problem and locate a way to solve it while others were still trying to understand the question. He was an extremely hard worker and persuaded others to work as hard as he did. Soon after arriving at Princeton, he became editor of the *Annals of Mathematical Statistics*, the journal started by Henry Carver. He raised the standards of publication and brought his graduate students into the editing of the journal. He persuaded John Tukey, a new faculty member with an initial interest in more abstract mathematics, to join him in statistical research. He took on a sequence of graduate students who went out to found or staff the new departments of statistics in many universities after World War II.

Wilks's initial thesis on a problem in educational psychology led him to work with the Educational Testing Service, where he helped formulate the sampling procedures and the scoring tech-

[1]There was another fellow at the institute named Albert Einstein. But he was a physicist and while his accomplishments were a little more complicated than Moore's "sweeping the streets," his work was badly tainted with applications to "real life."

niques used for college entrance and other professional school exams. His theoretical work established the degree to which weighted scoring schemes could differ and still produce similar results. He was in contact with Walter Shewhart[2] at Bell Telephone Laboratories, who was beginning to apply Fisher's experimental design theories to industrial quality control.

Statistics and the War Effort

As the 1940s approached, perhaps Wilks's most important work came from his consulting with the Office of Naval Research (ONR) in Washington. Wilks was convinced that experimental design methods could improve the weapons and the fire doctrines of the armed forces, and he found a receptive ear at the ONR. By the time the United States entered World War II, the army and the navy were both ready to apply statistical methods in the American version of operations research. Wilks set up the Statistical Research Group–Princeton (SRG-P) under the National Defense Research Council. The SRG-P recruited some of the brightest young mathematicians and statisticians, many of whom would make major contributions to the science in the years following the war. These included John Tukey (who swung over entirely to applications), Frederick Mosteller (who would go on to found the several statistics departments at Harvard), Theodore W. Anderson (whose textbook on multivariate statistics was to become the bible of that field), Alexander Mood (who was to go on to make major advances in the theory of stochastic processes), and Charles Winsor (who was to give his name to an entire class of estimation methods), among others.

[2]Today, almost every quality control department in industry uses Shewhart charts to track the variations in output. The name *Shewhart* is a partial example of Stigler's law of misonomy. The actual mathematical formulation of the Shewhart chart seems to have been first proposed by Gosset ("Student") and may even be seen in an early textbook by George Udny Yule. But Walter Shewhart showed how to apply this technique to quality control and popularized it as an effective methodology.

Richard Anderson, working at that time as a graduate student with the SRG-P, describes attempts that were being made to find a method of destroying land mines. As the invasion of Japan neared, the American army learned that the Japanese had developed a nonmetallic land mine that could not be detected by any known means. They were planting these mines in random patterns across the beaches of Japan and along any possible invasion routes. Estimates of deaths from those land mines alone ran into the hundreds of thousands. It was urgent to find a way to destroy them. Attempts to use bombs dropped from airplanes against land mines in Europe had failed. Anderson and others from the SRG-P were set to designing experiments on the use of lines of explosive cord to destroy the mines. According to Anderson, one of the reasons why the United States dropped the atomic bomb on Japan was that all their experiments and calculations showed that it was impossible to destroy these mines by such means.

The group worked on the effectiveness of proximity fuses in antiaircraft projectiles. A proximity fuse sends out radar signals and explodes when it is close to a target. The group helped develop the first of the smart bombs that could be steered toward their target. They worked on range finders and different types of explosives. Members of the SRG-P found themselves designing experiments and analyzing data at ordnance laboratories and army and navy facilities all over the country. Wilks helped organize a second group, called Statistical Research Group–Princeton, Junior (SRG-Pjr), at Columbia University. Out of SRG-Pjr came "sequential analysis." This was a means of modifying the design of an experiment while the experiment is still running. The modifications allowed by sequential analysis involve the very treatments being tested. Even in the most carefully designed experiment, it sometimes happens that the emerging results suggest that the original design should be changed to produce more complete results. The mathematics of sequential analysis allow the scientist to know which modifications can be made and which cannot, without affecting the validity of the conclusions.

The initial studies in sequential analysis were immediately declared top secret. None of the statisticians working on it were allowed to publish until several years after the war was over. Once the first papers on sequential analysis and its cousin, "sequential estimation," began to appear in the 1950s, the method captured the imagination of others, and the field rapidly developed. Today, sequential methods of statistical analysis are widely used in industrial quality control, in medical research, and in sociology.

Sequential analysis was just one of the many innovations that came out of Wilks's statistical research groups during World War II. After the war, Wilks continued to work with the armed forces, helping them improve the quality control of their equipment, using statistical methods to improve planning for future needs, and bringing statistical methods into all aspects of military doctrine. One of Wilks's objections to the mathematicians who continued to inhabit their world of pure abstractions was that they were not being patriotic. He felt the country needed the brainpower they were siphoning away into these purposely useless abstractions. This brainpower needed to be applied, first to the war effort and then to the Cold War afterward.

There is no record of anyone getting angry at Samuel S. Wilks. He approached everyone he dealt with, whether a new graduate student or a four-star general of the army, with the same informal air. He was nothing but an old Texas farm boy, he would imply, and he knew he had a lot to learn, but he wondered if. . . . What followed this would be a carefully reasoned analysis of the problem at hand.

Statistics in Abstraction

Sam Wilks struggled to make mathematical statistics both a respectable part of mathematics and a useful tool for applications. He tried to move his fellow mathematicians away from the cold world of abstraction for the sake of abstraction. There is indeed a fundamental beauty in mathematical abstractions. They so attracted the Greek philosopher Plato that he declared that all those things that we can see and touch are, in fact, mere shadows of the true reality and that the real things of this universe can be found only through the use of pure reason. Plato's knowledge of mathematics was relatively naive, and many of the cherished purities of Greek mathematics have been shown to be flawed. However, the beauty of what can be discovered with pure reason continues to entice.

In the years since Wilks was editor of the *Annals of Mathematical Statistics*, the articles that appear in the *Annals*[3] and in *Biometrika* have become more and more abstract. This has also been true of articles in the *Journal of the American Statistical Association* (whose early issues were devoted to descriptions of government statistical programs) and the *Journal of the Royal Statistical Society* (whose early issues contained articles listing detailed agricultural and economic statistics from throughout the British Empire).

The theories of mathematical statistics, once thought by mathematicians to be too mired in messy practical problems, have become clarified and honed into mathematical beauty. Abraham Wald unified the work on estimation theory by creating a highly abstract generalization known as "decision theory," wherein different theoretical properties produce different criteria for estimates. R. A. Fisher's work on the design of experiments made use of theorems from finite group theory and opened up fascinating ways of looking at comparisons of different treatments. From this came a branch of mathematics called "design of experiments," but the papers published in this field often deal with experiments so complicated that no practicing scientist would ever use them.

Finally, as others continued to examine the early work of Andrei Kolmogorov, the concepts of probability spaces and stochastic processes became more and more unified but more and more

[3]In the early 1980s, the rapid development of statistical theory caused the *Annals* to divide into two journals, the *Annals of Statistics* and the *Annals of Probability*.

abstract. By the 1960s, papers published in statistical journals dealt with infinite sets on which were imposed infinite unions and intersections forming "sigma fields" of sets—with sigma fields nested within sigma fields. The resulting infinite sequences converge at the point of infinity, and stochastic processes hurl through time into small bounded sets of states through which they are doomed to cycle to the end of time. The eschatology of mathematical statistics is as complicated as the eschatology of any religion, or more so. Furthermore, the conclusions of mathematical statistics are not only true but, unlike the truths of religion, they can be *proved* true.

In the 1980s, the mathematical statisticians awoke to a realization that their field had become too far divorced from real problems. To meet the urgent need for applications, universities began setting up departments of biostatistics, departments of epidemiology, and departments of applied statistics. Attempts were made to rectify this breaking apart of the once unified subject. Meetings of the Institute of Mathematical Statistics were devoted to "practical" problems. The *Journal of the American Statistical Association* set aside a section of every issue to deal with applications. One of the three journals of the Royal Statistical Society was named *Applied Statistics*.[4] Still, the siren calls of abstraction continued. The Biometric Society, set up in the 1950s, had introduced a journal named *Biometrics*, which would publish the applied papers that were no longer welcome in *Biometrika*. By the 1980s, *Biometrics* had become so abstract in its contents that other journals, like *Statistics in Medicine*, were created to meet the need for applied papers.

[4]Soon after World War II, the *Journal of the Royal Statistical Society* split into three journals, initially named *JRSS Series A*, *JRSS Series B*, and *JRSS Series C*. *Series C* eventually was named *Applied Statistics*. The Royal Statistical Society tries to keep *Series A* to deal with general issues affecting business and government. *Series B* is where mathematical statistics, with all its abstractions, can be found. It has been a struggle to keep *Applied Statistics* applied, and each issue has articles whose "applications" are quite far-fetched and appear to be there only to justify the development of yet another gem of beautiful, but abstract, mathematics.

The mathematics departments of American and European universities missed the boat when mathematical statistics arrived on the scene. With Wilks leading the way, many universities developed separate statistics departments. The mathematics departments missed the boat again when the digital computer arrived, disdaining it as a mere machine for doing engineering calculations. Separate computer science departments arose, some of them spun off from engineering departments, others spun off from statistics departments. The next big revolution that involved new mathematical ideas was the development of molecular biology in the 1980s. As we shall see in chapter 28, both the mathematics and the statistics departments missed that particular boat.

Samuel S. Wilks died at age fifty-eight in 1964. His many students have played major roles in the development of statistics during the last fifty years. His memory is honored by the American Statistical Association with the annual presentation of the S. S. Wilks Medal to someone who meets Wilks's standards of mathematical creativity and engagement in the "real world." The old Texas farm boy had made his mark.

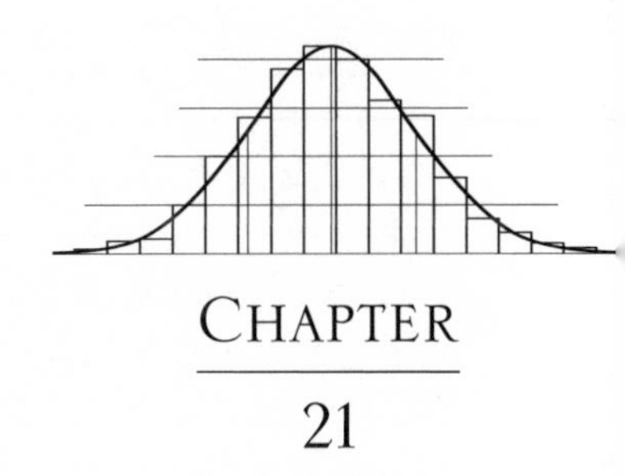

Chapter 21

A Genius in the Family

The first quarter of the twentieth century witnessed a mass migration from eastern and southern Europe to England, the United States, Australia, and South Africa. Most of these millions of immigrants were from the poorest classes of their homelands, seeking economic opportunity and freedom from oppressive rulers or chaotic governments. Many of them settled in the slums of big cities, where their children were urged to climb out of this squalor via the magic wand of education. Some of those children showed remarkable promise. Some were even geniuses. This is the story of two such children of immigrants, one of whom accumulated a Ph.D. and two Sc.D.s, the other of whom dropped out of high school at age fourteen.

I. J. Good

Moses Goodack had no love for the czar or for the czar's Polish domain, where he was born. In particular, he had no desire to be conscripted into the czar's army. When he was seventeen, he escaped to the West with a similarly minded friend. They had thirty-five rubles and a large cheese between them. Without railroad tickets, they slept beneath the seats of the trains they took and used the cheese to bribe the ticket takers. Goodack arrived in the Jewish slum district of Whitechapel, in London, with nothing but his pluck and his health. He opened a shop as a watchmaker, having learned his trade by looking at other watchmakers working in their show windows (where the light was best). He became interested in antique cameos and eventually opened a shop near the British Museum that specialized in antique jewelry (using money borrowed from his wife-to-be). The sign painter hired to paint the name of the new business on its window was too drunk to understand how to spell *Goodack*, so the business became "Good's Cameo Corner," and the family name became Good.

Moses Goodack's son, I. J. Good, was born in London on December 9, 1916. His given first name was Isidore. Young Isidore Good was embarrassed by posters that appeared all over town advertising a play entitled *The Virtuous Isidore*. From that time on, he was known as Jack, and he published his papers and books as I. J. Good.

In an interview with David Banks, in 1993, Jack Good recalled that, when he was about nine years old, he discovered numbers and became proficient at mental arithmetic. He had come down with diphtheria and had to remain in bed. One of his older sisters showed him how to extract a square root. At one time, extraction of square roots was taught as part of the regular school curriculum after the students had learned how to do long division. It involved a sequence of halving and squaring operations that were laid out on the paper in a fashion similar to long division.

Forced to lie quietly in bed, he began to work out the square root of two in his head. He discovered that the operation continued indefinitely, and when he squared his partial answer the square of that answer was just slightly less than two. He kept examining the numbers that emerged, looking for a pattern, but was unable to find one. He realized that the operation could be thought of as the difference between one square and twice another square. Because of this, it could be represented as the ratio of two numbers only if there was a pattern. Lying in bed, working entirely within his head, ten-year-old Jack Good discovered the irrationality of the square root of two. In the course of this, he also found a solution to a Diophantine problem known as "Pell's equation." The facts that the irrationality of the square root of two had been discovered by the Pythagorean Brotherhood in ancient Greece and that Pell's equation had been solved in the sixteenth century do not detract from this remarkable achievement of mental arithmetic in a ten-year-old boy.

In the 1993 interview, Jack Good mused, "It wasn't a bad discovery that Hardy [a British mathematician active in the 1920s and 1930s] described as one of the greatest achievements of the ancient Greek mathematicians. Being anticipated by great men is now familiar to me, but it was not usually by 2.5 millennia."

At age twelve, Good entered the Haberdashers' Aske's School[1] in Hampstead, a secondary school for boys whose motto is "Serve and obey." It catered to the children of shopkeepers and maintained rigorous standards. Only 10 percent of its students were able to make it to the "top form," the highest grade, and only a sixth of those went on to university education. Early in his career, Good was taught by a schoolmaster named Mr. Smart. Mr. Smart would often write a group of exercise questions on the blackboard; some of them were so devilishly difficult that he knew the students would have to spend time working them out, while he was able to do paperwork at his desk. As he finished writing the last question, young I. J. Good announced, "I've finished." "You mean with the first one?" Mr. Smart asked with some astonishment. "No," Jack replied. "All of them."

[1]The Haberdashers' Aske's School is one of seven schools founded by the Haberdashers' Company, an ancient livery company (incorporated in 1448). Robert Aske, Past Master of the Haberdashers' Company, died in 1689, leaving a bequest (to this day managed by the company) for the founding of a school for twenty sons of poor Haberdashers. Today it is a highly successful school of 1,300 boys. The connection with the Haberdashers' Company is retained through the governing body, of which half the members, including the chairman, are members of the company.

Good was fascinated with books of mathematical puzzles, but he preferred to look at the answers first and then find a way to get from the puzzle to the proposed answer. Confronted with a problem involving piles of shot, he looked at the answer and realized that it could have been arrived at by tedious calculation, but he was intrigued by the possibility of generalizing the answer. In doing so, he discovered the principle of mathematical induction. Good was improving. This was a discovery that had been made by earlier mathematicians only 300 years before.

At nineteen, Good entered Cambridge University, his reputation as a mathematical prodigy having preceded him. Here, however, he discovered many other students who were as brilliant as he. His math tutor at Jesus College, Cambridge, seemed to take delight in presenting proofs that had been cleaned up to such a degree of slickness that the intuitive ideas behind the proofs had been completely obliterated. To make things more difficult, he would present the proofs so rapidly that students had difficulty copying them down from the board before he erased the old lines to start new ones. Good excelled at Cambridge, catching the eye of some of the senior mathematicians at the university. In 1941, he emerged with a Ph.D. in mathematics. His thesis dealt with the topological concept of partial dimension, which was an extension of ideas developed by Henri Lebesgue (recall the Lebesgue whose work Jerzy Neyman had admired but who was so rude to young Neyman when they met).

There was a war on, and Jack Good became a cryptanalyst in the laboratories at Bletchley Park, near London, where attempts were being made to break the German secret codes. A secret code consists of converting the letters of the message into a sequence of symbols or numbers. By 1940, these codes had become very complicated, with the pattern of conversion changing with each letter. Suppose, for instance, you want to encrypt the message "War has begun." One way is to assign numbers to each letter, so the encryption becomes: "12 06 14 09 06 23 11 19 20 01 13." The cryptanalyst would notice the multiple occurrence of the number 06 and conclude that this was a repetition of the same letter. With a long enough message, and some knowledge of the statistical frequency of different letters in the language being encrypted, along with a few lucky guesses, the cryptanalyst can usually decode such a message in a few hours.

The Germans developed a machine during the last years of World War I that changed the code with each letter. The first letter might be encrypted with 12, but before encrypting the next letter, the machine would pick an entirely different code, so the second letter might be encoded with 14, with a change at the next letter, and so on. In this way, the cryptanalyst cannot depend upon repeated numbers as indicating the same letters. This new type of code has to be understood by its intended recipient, so there has to be some degree of regularity to the way in which the machine changes from one code to another. The cryptanalyst can look at statistical patterns and estimate the nature of the regularity and crack the code that way. The task can be made even harder for the cryptanalyst: Once the initial codes can be shifted by a fixed plan, it is possible to shift the plan by a superfixed plan, and the code becomes even more difficult to break.

All of this can be represented in a mathematical model that resembles the hierarchal Bayes models of chapter 13. The pattern of shifting at each level of encryption can be represented by parameters; so we have the case of measurements, the initial numbers in the coded messages observed, parameters that describe the first level of coding, hyperparameters that describe the changes in these parameters, hyper-hyperparameters that describe the changes in these hyperparameters, and so forth. Eventually, because the code has to be broken by the recipient, there has to be a final level of the hierarchy where the parameters are fixed and unchangeable, and so all such codes are theoretically breakable.

One of I. J. Good's major achievements was the contribution he made to the development of empirical Bayes and hierarchal

Bayes methods which he derived from the work he did at Bletchley. He emerged from his wartime experiences with a deep interest in the underlying theories of mathematical statistics. He taught briefly at the University of Manchester, but the British government lured him back into intelligence, where he became an important figure in the adoption of computers to deal with problems of cryptanalysis. The power of the computer to examine vast numbers of possible combinations led him into investigations of classification theory, where observed units are organized in terms of different definitions of "closeness."

While working with British Intelligence, Good picked up two more advanced degrees, a Sc.D. from Cambridge and a D.Sc. from Oxford. He came to the United States in 1967, where he accepted a position as University Distinguished Professor at Virginia Polytechnic Institute and stayed there until his retirement in 1994.

Good was always intrigued by apparent coincidences in the occurrence of numbers. "I arrived in Blacksburg [Virginia] in the seventh hour of the seventh day of the seventh month of the year seven in the seventh decade, and I was put in apartment 7 of block 7 . . . all by chance." He continued, whimsically, "I have a quarter-baked idea that God provides more coincidences the more one doubts Her existence, thereby providing one with evidence without forcing one to believe." This eye for coincidences led to work in statistical estimation theory. Because the human eye is capable of seeing patterns in purely random numbers, at what point, he asked, does an apparent pattern really mean something? Good's mind probed the underlying meaning of the models used in mathematical statistics, and his later papers and books tend to become more and more philosophical.

PERSI DIACONIS

An entirely different career awaited Persi Diaconis, born to Greek immigrants in New York, on January 31, 1945. Like I. J. Good, young Persi was intrigued with mathematical puzzles. Whereas Good read the books of H. E. Dudeney, whose mathematical puzzles entertained Victorian England, Diaconis read the "Mathematical Recreations" column by Martin Gardner in *Scientific American*. Eventually, while still in high school, Diaconis met Martin Gardner. Gardner's column often dealt with tricks of cardsharps and methods used to make things appear to be different, and these interested Diaconis, especially when they involved intricate problems in probability.

So intrigued was Persi Diaconis with cards and tricks that he ran away from home at age fourteen. He had been doing magic tricks since he was five years old. In New York, he frequented the shops and restaurants where other magicians gathered. At one cafeteria he met Dia Vernon, who traveled around the United States with a magic show. Vernon invited Persi to join his show as an assistant. "I jumped at the chance," Diaconis related. "I just went off. I didn't tell my parents; I just left."

Vernon was in his sixties at the time. Diaconis traveled with him for two years, learning Vernon's repertoire of tricks and devices. Then Vernon left the road to settle down with a magic shop in Los Angeles, and Diaconis continued a traveling magic show of his own. People had trouble pronouncing his name, so he took as his professional name Persi Warren. As he recounted:

> That's not a great life, but it's OK. Working in the Catskills. What happens is that you get somebody who sees you and likes you and says, "Gee, would you ever think of coming to Boston? . . . I'll pay your fare and $200." . . . And you go to Boston . . . and you check into a show business rooming house. You do that date and maybe an agent will get you another job while you're there, and so forth.

At age twenty-four, Persi Diaconis returned to New York, tired of the traveling magician game. He did not have a high school diploma. He had been accelerated in school, and, when he dropped out at age fourteen, he had less than a year to complete his high school education. Without a diploma, he enrolled in the general studies program at City College of New York (CCNY). He found that over the years he had been away from home, he had been sent a great deal of mail from the army and from different technical institutes and colleges. They were all form letters inviting him to join their institutions, and they all began, "Dear Graduate." It turned out that when he ran away from home and school, his teachers decided to put him up for graduation anyway, giving him final grades for the courses he was taking, which would have been adequate for him to graduate. Unknown to young Persi, he was officially a graduate of George Washington High School in New York.

He went to college for a strange reason. He had bought a copy of a graduate level textbook on probability theory by William Feller of Princeton University. He had difficulty understanding it (as do most people who try to make their way through Feller's tightly written *Introduction to Probability Theory and Its Applications*, Vol. I[2]). Diaconis entered CCNY in order to learn enough formal mathematics to understand Feller. In 1971, at age twenty-six, he received his undergraduate degree from CCNY.

He was accepted by a number of different universities for graduate studies in mathematics. He had been told that no one ever got into the Harvard mathematics department from CCNY (which was not true), so he decided to apply to the Harvard statistics department instead of the mathematics department. He wanted to go to Harvard and he thought, as he said later, that if he didn't like statistics, "Well, I'll transfer to math or something. They'll learn that I'm terrific . . ." and would accept the transfer. Statistics did

[2]There is a perverse element in the titles of mathematical texts. The most difficult are usually entitled *Introduction to* . . . or *The Elementary Theory of* . . . Feller's book is doubly difficult, since it is an "introduction" and it is only volume I.

interest him; he got his Ph.D. in mathematical statistics in 1974 and took a position at Stanford University, where he rose to the position of full professor. As of this writing, he is a professor at Harvard University.

The electronic computer has completely restructured the nature of statistical analysis. At first, the computer was used to do the same sort of analyses that had been done by Fisher, Yates, and others, only much faster and with much greater ambition. Recall (from chapter 17) the difficulty Jerry Cornfield had when it was necessary to invert a 24 × 24 matrix. Today, the computer that sits on my desk can invert matrices on the order of 100 × 100 (although anyone who gets stuck with such a situation has probably not done a good job of defining the problem). Matrices that are ill-conditioned can be operated on to produce generalized inverses, a concept that was purely theoretical in the 1950s. Large, complicated analyses of variances can be run on data generated by experimental designs that involve multiple treatments and cross-indexing. This sort of effort involves mathematical models and statistical concepts that go back to the 1920s and 1930s. Can the computer be used for something different?

In the 1970s, at Stanford University, Diaconis joined a group of other young statisticians who were looking at the structure of the computer and of mathematical statistics, asking that very question. One of the earlier answers was a method of data analysis known as "projection pursuit." One of the curses of the modern computer is that it is possible to assemble data sets of huge dimension. Suppose, for instance, we are following a group of patients who have been diagnosed at high risk for heart disease. We bring them into the clinic for observation once every six months. At each visit, we take 10 cc of blood and analyze it for the levels of over 100 different enzymes, many of which are thought to be related to heart disease. We also subject the patients to echocardiograms, producing about six different measures, and to EKG monitoring (perhaps having them wear a device that records all 900,000 heart beats in a given day). They are measured, and weighed, and poked, and probed for clinical signs and symptoms, resulting in another thirty to forty measurements.

What can be done with all this data?

Suppose we produce 500 measurements at each clinic visit for each patient, and that the patients are seen at 10 different visits over the course of the study. Each patient provides us with 5,000 measurements. If there are 20,000 patients in the study, this can be represented as 20,000 points in a 5,000-dimensional space. Forget the idea of travel through a mere four dimensions that preoccupies much of science fiction. In the real world of statistical analysis, it is not unusual to have to deal with a space of thousands of dimensions. In the 1950s, Richard Bellman produced a set of theorems, which he called "the curse of dimensionality." What these theorems say is that, as the dimension of the space increases, it becomes less and less possible to get good estimates of parameters. Once the analyst is out in ten- to twenty-dimensional space, it is not possible to detect anything with less than hundreds of thousands of observations.

Bellman's theorems were based on the standard methodology of statistical analysis. What the Stanford group realized was that real data are not scattered helter-skelter in this 5,000-dimensional space. The data actually tend to bunch up into lower dimensions. Imagine a scatter of points in three dimensions that really all lie on a single plane or even on a single line. This is what happens to real data. The 5,000 observations on each patient in the clinical study are not scattered about without some structure. This is because so many of the measurements are correlated with each other. (John Tukey from Princeton and Bell Labs once proposed that, at least in medicine, the true "dimensionality" of data is often no more than five.) Working with this insight, the Stanford group developed computer-intensive techniques for searching out the lower dimension that is really there. The most widely used of these techniques is projection pursuit.

In the meantime, this proliferation of information in large, ill-structured masses has attracted the interest of other scientists, and the field of information science has emerged at many universities. In many cases, these information scientists have engineering training and are unaware of many of the recent developments of mathematical statistics, so there has been a parallel development in the world of computer science that sometimes rediscovers the material of statistics and sometimes opens up new doors that were never anticipated by R. A. Fisher or his followers. This is a topic of the final chapter of this book.

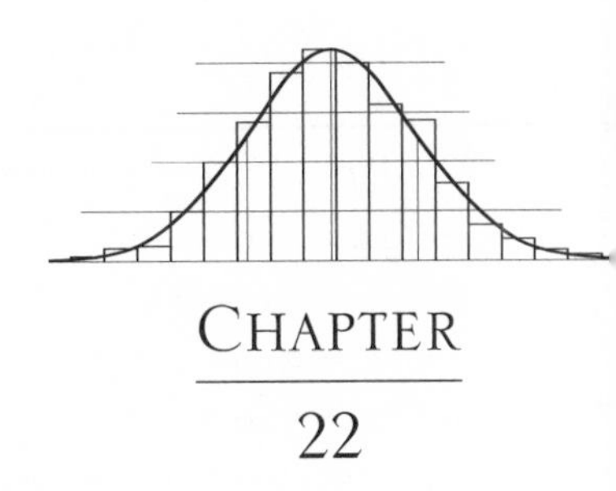

CHAPTER

22

THE PICASSO OF STATISTICS

When I completed my doctoral thesis in 1966, I made the rounds of different universities, giving talks about my results and allowing myself to be interviewed for a possible position. One of my first stops was Princeton University, where I was met at the train station by John Tukey.

In my studies, I had learned about Tukey contrasts, Tukey's one degree of freedom for interaction, Tukey's fast Fourier transforms, Tukey's quick test, and Tukey's lemma. I had not yet learned about his work in exploratory data analysis or any of the work that was to flow from his fecund mind in the coming years. John Tukey was chairman of the Department of Statistics (and also held a joint appointment at Bell Telephone Laboratories), and it was surprising to me that he would personally greet my arrival. He was dressed in cotton chino pants with a loose sport shirt and tennis shoes or sneakers. I was dressed in a suit and tie. The sartorial revolution of the sixties had not yet risen to the faculty level, so my style of dress was more appropriate than his.

Tukey walked with me across campus. We discussed living conditions at Princeton. He inquired about the computer programs I had written while working on my thesis. He showed me some tricks to avoid round-off error in my programs. We finally arrived at the hall where I was to give a talk on my thesis. After introducing me, Tukey climbed up to the last row of seats in the room. I began to present my results. As I did so, I noticed that he was busy correcting papers.

When I finished my presentation, several of the people in the audience (which consisted of graduate students and faculty members) asked some questions or suggested ramifications of what I had done. When it was clear that no one else had comments or questions, John Tukey walked down from the back row. He took a piece of chalk and reproduced my main theorem on the blackboard, complete with all my notation.[1] He then showed me an alternative proof of the theorem it had taken me months to prove in my own way. "Wow," I said to myself. "This is the big leagues!"

John Tukey was born in 1915 in New Bedford, Massachusetts. The lingering sound of his near-Boston accent salted his speech. His parents recognized his genius early, and he was taught at home until he left to attend Brown University, where he earned his bachelor's and master's degrees in chemistry. He had become intrigued

[1]Mathematical notation, consisting of an array of letters, both Roman and Greek, with squiggly lines, and superscripts and subscripts, is one aspect of mathematics that intimidates the nonmathematician (and often some mathematicians, too). It is really a convenient way to relate complicated ideas to each other in a compact space. The "trick" in reading a mathematical paper is to recognize that each symbol has some meaning, to know the meaning when it is introduced, but then to take it on faith that you "understand" its meaning, and to pay attention to the way in which the symbol is manipulated. The essence of mathematical elegance is to produce a notation of symbols that is so simply organized that the reader can understand the relationships immediately. This is the sort of elegance one finds in papers by Jerzy Neyman. My Ph.D. thesis, I fear, was far from elegant. I was using the notation to make sure that all possible aspects of the mathematical model were included. My subscripts had subscripts and my superscripts had subscripts, and, in some cases, the subscripts had variable arguments. I was astonished that John Tukey was able to reproduce this complicated mess of symbols completely out of his head, having seen the theorem for the first time that afternoon. (In spite of the messiness of my notation, Tukey did offer me a position. But I had three children and one on the way, and I accepted a higher-paying position elsewhere.)

by abstract mathematics. He continued his graduate work at Princeton University, where he received his doctorate in mathematics in 1939. His initial work was in the field of topology. Point set topology provides an underlying theoretical foundation for mathematics. Beneath the topological foundation is a difficult and arcane branch of philosophy known as "metamathematics." Metamathematics deals with questions of what it means to "solve" a mathematical problem, and what are the unstated assumptions behind the use of logic. Tukey plunged into these murky foundations and emerged with "Tukey's lemma," a major contribution to the field.

John Tukey was not destined to remain in abstract mathematics, however. Samuel S. Wilks was on the Princeton mathematics faculty, pushing students and young faculty members into the world of mathematical statistics. Upon receiving his doctorate, Tukey was named an instructor in the Department of Mathematics at the university. In 1938, while still working on his thesis, he published his first paper in mathematical statistics. By 1944, almost all of his publications were in that field.

During World War II, Tukey joined the Fire Control Research Office, working on problems of aiming guns, evaluating ranging instruments, and related problems of ordnance. This experience provided him with many examples of statistical problems that he would investigate in future years. It also gave him a great appreciation for the nature of practical problems. He is noted for his one-line aphorisms that sum up important experiences. One of these, emerging from his practical work, is "It is better to have an approximate answer to the right question than an exact answer to the wrong one."

Tukey's Versatility

Pablo Picasso astonished the world of art in the early years of the twentieth century with his protean output. For a while, he played with monochromatic paintings, then he invented cubism, then he examined a form of classicism, then he went on to ceramics. Each of these excursions resulted in a revolutionary change in art, which others continued to exploit after Picasso went on to other things. So it has been with John Tukey. In the 1950s, he became involved with Andrei Kolmogorov's ideas on stochastic processes and invented a computer-based technique for analyzing long strings of correlated results, known as the "fast Fourier transform," or FFT. Like Picasso and cubism, Tukey could have accomplished nothing more and his influence on science would have been immense.

In 1945, Tukey's war work took him to Bell Telephone Laboratories in Murray Hill, New Jersey, where he became involved in different practical problems. "We had an engineer named Budenbom," he said in a conversation recorded in 1987, "who had been building a new, especially good, tracking radar for tracking aerial targets. He wanted to go to California to give a paper and he wanted a picture to show what his tracking errors were like." Budenbom had formulated his problem in the frequency domain but did not know how to get consistent estimates of the frequency amplitudes. Although Tukey, as a mathematician, was familiar with Fourier transformations, he had not yet been exposed to the uses of this technique in engineering. He proposed a method that seemed to satisfy the engineer (recall Tukey's aphorism about the usefulness of an approximate answer to the right question). Tukey himself was not satisfied, however, and continued thinking about the problem.

The result was the fast Fourier transform. It is a smoothing technique that, to use Tukey's expression, "borrows strength" from neighboring frequencies, so vast amounts of data are not needed to get good estimates. It is also a carefully thought-out theoretical solution with optimum properties. Finally, it is a highly efficient computer algorithm. Such an algorithm was needed in the 1950s and 1960s, when computers were much slower and had smaller memories. It continues to be used into the twenty-first century, because it is superior in its accuracy to some more complicated transformation estimates that can now be done.

The computer and its capacities have continually been advancing the frontiers of statistical research. We have seen previously how the computer made it possible to compute inverses of large matrices (something that would have taken Jerry Cornfield hundreds of years of labor on a mechanical calculator). There is another aspect of the computer that threatens to overwhelm statistical theory. This is the capacity of the computer to store and analyze huge amounts of data.

In the 1960s and early 1970s, engineers and statisticians at Bell Telephone Laboratories pioneered the analysis of huge amounts of data. Monitoring telephone lines for random errors and problems led to millions and millions of data items in a single computer file. Data being telemetered from space probes sent to examine Mars and Jupiter and other planets produced other files of millions of items. How can you look at so much data? How do you begin to structure it so it can be examined?

We can always estimate the parameters of the probability distributions, following the techniques pioneered by Karl Pearson. This requires us to assume something about those distributions—that they belong to the Pearson system, for instance. Can we find methods for examining these vast collections of numbers and learn something from them without imposing assumptions about their distributions? In some sense, this is what good scientists have always done. Gregor Mendel ran a series of plant crossing experiments, examining the output and gradually developing his theories of dominant and recessive genes. Although a great deal of scientific research involves collecting data and fitting those data to the preconceived mold of some specific type of distribution, it is often useful and important just to collect data and examine them carefully for unexpected events.

As Eric Temple Bell, the American mathematician, once pointed out, "Numbers do not lie, but they have the propensity to tell the truth with intent to deceive."[2] The human being is prone to seeing patterns and will often see patterns where only random noise exists.[3]

This is particularly disturbing in epidemiology, when an examination of data often finds a "cluster" of diseases in a certain place or time. Suppose we find an unusually high number of children with leukemia in a small town in Massachusetts. Does this mean that there is some cause of cancer operating in this town? Or is this just a random cluster that happened to appear here and could just as easily have appeared elsewhere? Suppose the townspeople discover that a chemical plant has been dumping waste chemicals into a pond in a nearby town. Suppose they also find traces of aromatic amines in the water supply of the town where the cluster of leukemia occurred. Is this a possible cause of the leukemia? In a more general sense, to what extent can we examine data with an eye toward patterns and expect to find anything more than a random will-o'-the-wisp?

In the 1960s, John Tukey began to give serious consideration to these problems. He emerged from this with a highly refined version of Karl Pearson's approach to data. He realized that the distribution of observed data can be examined as a distribution, without imposing some arbitrary probability model on it. The result

---

[2]Bell wrote several popular books on mathematics during the 1940s and 1950s. His *Men of Mathematics* is still the classic biographical reference for the great mathematicians of the eighteenth and nineteenth centuries. His *Numerology*, from which this quote was taken, deals with numerology, to which he was introduced, he writes, by his cleaning lady.

[3]A classic example of this is Bode's law. Bode's law is the empirical observation that there is a linear relationship between the logarithm of the distance from the Sun and the number of the planets in our solar system. In fact, the planet Neptune was found because astronomers applied Bode's law to predict the approximate orbit of another planet and found Neptune in that orbit. Until space probes of Jupiter and Saturn discovered many smaller moons close to the planets, the only observed satellites of Jupiter appeared to follow Bode's law. Is Bode's law a random coincidence? Or does it tell us something deep and not yet understood about the relationships between planets and the Sun?

was a series of papers, talks at meetings, and finally books dealing with what he called "exploratory data analysis." As he worked on the problems, Tukey's mode of presentation took on a strikingly original form. To shock his listeners and readers into reexamining their assumptions, he began to give different names to the characteristics of data distributions than had been used in the past. He also moved away from standard probability distributions as the starting point of analysis and toward examination of the patterns within the data themselves. He looked at the way in which extreme values can influence our observations of patterns. To adjust for such false impressions, he developed a group of graphical tools for displaying data.

For instance, he showed that the standard histograms for displaying the distribution of a set of data tend to be misleading. They lead the eye to looking at the most frequent class of observations. He proposed that, instead of plotting the frequency of observations, we plot the *square root* of the frequency. He called this a "rootgram," as opposed to a histogram. Tukey also proposed that the central region of data be plotted as a box and that extreme values be plotted as lines (or, as he called them, "whiskers") emanating from the box. Some of the tools he proposed have become part of many standard statistical software packages, and the analyst can now call for "box plots" and "stem and leaf plots." Tukey's fertile imagination has swept through the landscape of data analysis, and many of his proposals have yet to be incorporated into computer software. Two of his inventions have been incorporated into the English language. It was Tukey who coined the words *bit* (for "*bi*nary dig*it*") and *software* (computer programs, as opposed to "hardware"—the computer).

Nothing has been too mundane for Tukey to attack with original insight, and nothing is too sacrosanct for him to question. Take the simple process of tallying. Most readers have probably been exposed to the use of a tally figure when counting something. The usual one, presented to us by generations of teachers, is to make four vertical marks and a fifth one crossing out the four. How many cartoons has the reader seen where a ragged prisoner has chalked a series of these tally marks on the wall of his prison?

It is a foolish way to tally, said John Tukey. Consider how easy it is to make a mistake. You might put the cross over three instead of four lines, or you might put down five lines and then a cross. The incorrect tally is hard to spot. It looks like a correct one unless you carefully examine the number of vertical lines. It makes more sense to use a tally mark where errors can be easily spotted. Tukey proposed a ten-mark tally. You first mark four dots to make the corners of a box. You then connect the dots with four lines, completing the box. Finally, you make two diagonal marks forming a cross within the box.

These examples, the fast Fourier transform, and exploratory data analysis, are just part of Tukey's enormous output. Like Picasso going from cubism, to classicism, to ceramics, to fabrics, John Tukey marched across the statistical landscape of the second half of the twentieth century, from time series, to linear models, to generalizations of some of Fisher's forgotten work, to robust estimation, to exploratory data analysis. From the deep theory of metamathematics, he emerged to consider practical problems, and finally to consider the unstructured evaluation of data. Wherever he put his mark, statistics was no longer the same. To the very day he died in the summer of 2000, John Tukey was still challenging his friends and coworkers with new ideas and questions about the old.

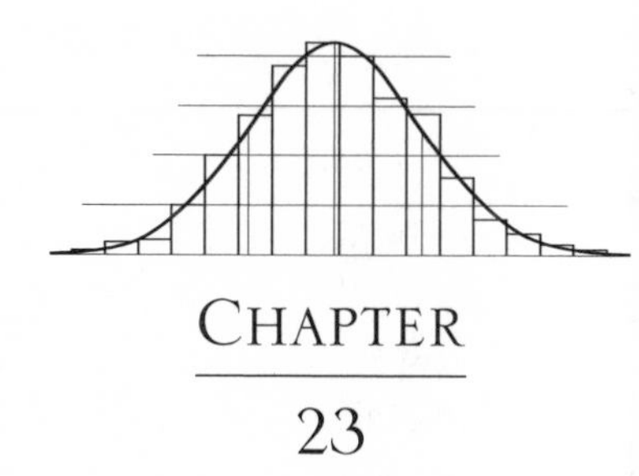

CHAPTER

23

DEALING WITH CONTAMINATION

The mathematical theorems that justify the use of statistical methods usually assume that the measurements made in a scientific experiment or observations are all equally valid. If the analyst picks through the data, using only those numbers that appear to be correct, the resulting statistical analysis can be in serious error. Of course, this is just what scientists have often done. In the early 1980s, Stephen Stigler went through the notebooks of great scientists of the eighteenth and nineteenth centuries, like Albert Michelson, who received the Nobel Prize in 1907 for determining the speed of light. Stigler found that all of them had discarded some of their data before beginning their calculations. Johannes Kepler, who discovered in the early seventeenth century that the planets orbited the Sun in ellipses, did so by reviewing the records of astronomers going back to some of the ancient Greeks; he often found that some observed positions did not fit the ellipses he was computing—so he ignored those faulty values.

But respectable scientists no longer throw out data that appear to be wrong. So extensive has the statistical revolution in science been that experimental scientists are now taught not to discard any of their data. The mathematical theorems of statistics require that all measurements be considered equally valid. What should be done when some of the measurements are obviously wrong? One day, in 1972, a pharmacologist came to my office with just such a problem. He was comparing two treatments for the prevention of ulcers in rats. He was sure that the treatments were producing different results, and his data seemed to tell him so. But when he ran a formal test of hypothesis following the Neyman-Pearson formulation, the comparison was not significant. The problem, he was sure, came from the data of two rats who had been treated with the poorer of the two compounds. Neither had any ulcers, making their results much better than the best result on the other treatment. We have seen in chapter 16 how nonparametric methods were developed to deal with this kind of problem. These two outlying values were on the wrong side of the data, and there were two of them, making even the nonparametric tests nonsignificant.

Had this happened 100 years ago, the pharmacologist would have thrown out the two wrong values and proceeded with his calculations. No one would have objected. He had been trained in the modern statistical approach to measurements, though. He knew that he was not allowed to do this. Fortunately, a new book was lying on my desk, which I had just finished studying. Entitled *Robust Estimates of Location: Survey and Advances*, it described a massive computer-oriented effort known as the Princeton Robustness Study, under the guidance of John Tukey. The answer to his problem was in that book.

"Robust," in this context, may seem a strange word to American ears. A great deal of the terminology of statistics comes from British statisticians and often reflects their use of the language. For instance, it is common to refer to the slight random fluctuations in data as "errors."[1] Sometimes, the data include values that are not only obviously wrong but for which it is possible to identify the reason why they are wrong (such as the complete failure of crops to grow in a given plot of land). Such data were called "blunders" by R. A. Fisher.

So it was that George Box, Fisher's son-in-law, produced the word "robust" out of his own British usage. Box has a strong accent that reflects his origins growing up near the mouth of the Thames. His grandfather had been a hardware merchant, and the business did well enough to send Box's older uncles on to university educations; one of them became a professor of theology. By the time Box's father reached manhood, however, the business had failed, and he went without higher education, trying to raise a family on the salary of a shopkeeper's assistant. George Box attended grammar school, and, knowing that there would never be enough money to go to a university, he began studying chemistry at a polytechnic institute. At that point, World War II began, and Box was drafted into the army.

Because he had been studying chemistry, he was sent to work at the chemical defense experiment station, where some of the leading pharmacologists and physiologists in the United Kingdom were trying to find antidotes for different poison gases. Among these scientists was Sir John Gaddum, who had brought the statistical

[1]There is often a confusion between the ordinary meaning of words and the specific mathematical meaning when the same words appear in a statistical analysis. When I first began to work in the drug industry, one of my analyses included a traditional table of results, where one line referred to the uncertainty produced by small random fluctuations in the data. This line is called, in the traditional table, the "error." One of the senior executives refused to send such a report to the U.S. Food and Drug Administration. "How can we admit to having error in our data?" he asked, referring to the extensive efforts that had been made to be sure the clinical data were correct. I pointed out that this was the traditional name for that line. He insisted that I find some other way to describe it. He would not send a report admitting error to the FDA. I contacted H. F. Smith at the University of Connecticut and explained my problem. He suggested that I call the line the "residual," pointing out that in some papers this is referred to as the residual error. I mentioned this to other statisticians working in the industry, and they began to use it. This eventually became the standard terminology in most of the medical literature. It seems that no one, in the United States at least, will admit to having error.

revolution into the science of pharmacology in the late 1920s, and who had put the basic concepts on a firm mathematical footing.

BOX BECOMES A STATISTICIAN

The colonel under whom George Box was working was puzzling over a large amount of data that had been accumulated on the effects of different poison gases at different doses on mice and rats. He could not make any sense out of them. As Box recounted in 1986:

> I said [to the colonel] one day, "You know, we really need to have a statistician look at these data because they are very variable." And he said, "Yes, I know, but we can't get a statistician; there isn't one available. What do you know about it?" I said, "Well, I don't know anything much about it, but I once tried to read a book called *Statistical Methods for Research Workers* by a man called R. A. Fisher. I didn't understand it, but I think I understand what he was trying to do." And he said, "Well, if you read the book you'd better do it."

Box contacted the education corps of the army to inquire about taking a correspondence course in statistical methods. There were no such courses available. The methods of statistical analysis had not yet moved into the standard university curriculum. Instead, he was sent a reading list. The reading list was nothing more than a collection of the books published up to that date. It included two books by Fisher, a book on statistical methods in educational research, another book on medical statistics, and one dealing with forestry and range management.

Box became intrigued with Fisher's ideas on experimental design. He found specific designs worked out in the forestry book and adapted these designs to the animal experiments. (Cochran and Cox's book, with its large number of carefully described designs, had not yet been published.) Often the designs were not quite appropriate, so using Fisher's general descriptions and building on what he had found, Box produced his own designs. One experiment that was most frustrating occurred when volunteers were asked to expose small patches of skin on each arm to mustard gas and different treatments. Each subject's two arms were correlated, and something had to be done to account for that in the analysis. Something had to be done, but there was nothing like this in the forestry book. Nor had Fisher discussed it in his book. Working from the fundamental mathematical principles, Box, the man whose only education had been an incomplete course in chemistry at a polytechnic school, derived the appropriate design.

Some idea of the power of Box's designs can be seen in a negative result. An American ophthalmologist arrived at the experimental station with what he knew to be the perfect antidote for the effects of lewisite, a small drop of which could blind a person. He had done numerous experiments on rabbits in the United States, and his sheaves of papers proved that this was the perfect answer. Of course, he knew nothing about Fisher's design of experiments. In fact, his experiments were all badly flawed: The effects of treatment could not be disentangled from the effects of extraneous factors scattered helter-skelter in his designs. The fact that rabbits have two eyes enabled Box to propose a simple experiment that made use of his new design for correlated blocks. This experiment quickly showed that the proposed antidote was useless.

A report was prepared to describe these results. The author was a major in the British army, and Sergeant Box wrote the statistical appendix, which explained how the results were obtained. The officials who had to approve the report insisted that the Box appendix be removed. It was much too complicated for anyone to understand.

(That is, the reviewers could not understand it.) But Sir John Gaddum had read the original report. He came by to congratulate Box on his appendix and learned that it was to be cut out of the final report. Dragging Box behind him, Gaddum stormed into the main hut of the complex and marched into a meeting of the committee in charge of reviewing reports. To use Box's words: "I felt very embarrassed. Here's this very distinguished guy, who reads the riot act to all these civil servants and says, 'Put the bloody thing back in.'" And they quickly did.

When the war ended, Box decided it would be worth his while to study statistics. Because he had read Fisher, and knew that Fisher was teaching at University College of the University of London he sought out the university. What he did not know was that Fisher had left London in 1943 to chair the Department of Genetics at Cambridge. Box found himself being interviewed by Egon Pearson, who had suffered some of Fisher's acid disdain for his work with Neyman on hypothesis testing. Box launched into an excited description of Fisher's work, explaining what he had learned about experimental design. Pearson listened quietly and then said, "Well, by all means you can come. But I think you'll learn there were one or two other people besides Fisher in statistics."

Box studied at University College, took his bachelor's degree, and then went on to study for a master's degree. He presented some of his work in experimental design and was told that it was good enough for a doctoral thesis, so he took his Ph.D. instead. At that point, Imperial Chemicals Industry (ICI) was the major company in Great Britain in the discovery of new chemicals and drugs. Box was invited to join their mathematical services group. He worked at ICI from 1948 to 1956, turning out a remarkable series of papers (often with coauthors) that extended the techniques of experimental design, examined methods for gradually adjusting the output of a manufacturing process to improve its yield, and represented the beginnings of his later work on the practical applications of Kolmogorov's theories of stochastic processes.

BOX IN THE UNITED STATES

George Box came to Princeton University to be the director of its Statistical Techniques Research Group and then went on to found a department of statistics at the University of Wisconsin. He has been honored by being named a fellow of all the major statistical organizations, has received several prestigious prizes for his accomplishments, and continues to be active in research and organizations since his retirement. His accomplishments range across many areas of statistical research, dealing with both theory and applications.

Box got to know Fisher in his days working for ICI, but never on a close personal basis. When he was directing the Statistical Techniques Research Group at Princeton, one of Fisher's daughters, Joan, was given the opportunity to go to the United States, where some friends found a position for her as a secretary at Princeton. They met and married, and Joan Fisher Box published a definitive biography of her father and his work in 1978.

One of Box's contributions to statistics was the word *robust*. He was concerned that many statistical methods relied on mathematical theorems that contained assumptions about the distributional properties of the data that might not be true. Could useful methods be found even if the conditions for the theorem do not hold? Box proposed calling such methods "robust." He made some initial mathematical investigations but decided that the concept of robustness was too vague. He opposed attempts to give it a more solid meaning, because he thought there was some advantage to having a general, vague idea to guide the choice of procedures. However, the idea took a life of its own. The robustness of a hypothesis test was defined in terms of the probability of error. By extending one of Fisher's geometric ideas, Bradley Efron of Stanford University proved in 1968 that "Student"'s t-test was robust in this sense. E. J. G. Pitman's methods were used to show that most of the nonparametric tests were robust in the same sense.

Then, in the late 1960s, John Tukey at Princeton and a group of his fellow faculty members and students attacked the problem of what to do with measurements that are apparently wrong. The result of this was the Princeton Robustness Study, published in 1972. The basic idea behind this study is that of a contaminated distribution. The measurements taken are assumed to come, most of them, from the probability distribution whose parameters we wish to estimate. But the measurements have been contaminated by a few values that come from another distribution.

A classic example of a contaminated distribution occurred during World War II. The U.S. Navy had developed a new optical range finder, which required the user to view a three-dimensional stereoscopic image of the target and place a large triangle "on top" of the target. An attempt was made to determine the degree of statistical error in this instrument by having several hundred sailors range on a target that was a known distance away. Before each sailor looked through the range finder, the position was reset according to a table of random numbers, so the previous setting would not influence him.

The engineers who designed the study did not know that 20 percent of humans cannot see stereoscopically. They have what is known as lazy eye. About one-fifth of the measurements taken on the range finder were completely wrong. With only the data from the study in hand, there was no way of knowing which came from sailors with lazy eye, so the individual measurements from the contaminating distribution could not be identified.

The Princeton study modeled a large number of contaminated distributions in a gigantic Monte Carlo[2] study on a computer. They were looking at methods of estimating the central tendency of a distribution. One thing they learned was that the much beloved

[2]In a Monte Carlo study, individual measurements are generated using random numbers to mimic the actual event that might occur. This is done many thousands of times over, and the resulting measurements are run through a statistical analysis to determine the effects of specific statistical methods on the situation being mimicked. The name is derived from that of the famous gambling casino in Monaco.

average is a very poor measure when the data have been contaminated. A classic example of such a situation was an attempt by Yale University in the 1950s to estimate the income of its alumni ten years after graduation. If they took the average of all the incomes, it would be quite high, since a very small number of the alumni were multimillionaires. In fact, over 80 percent of the alumni had incomes less than the average.

The Princeton Robustness Study discovered that the average was heavily influenced by even a single outlier from the contaminating distribution. This is what was happening to the data the pharmacologist brought me from his rat ulcer study. The statistical methods he had been taught to use all depended upon the average. The reader may protest: Suppose these extreme and seemingly wrong measurements are true; suppose they come from the distribution we are examining and not from a contaminating distribution. Throwing them out will only bias the conclusions.

The Princeton Robustness Study sought a solution that would do two things:

1. Downgrade the influence of contaminating measurements if they are there.

2. Produce correct answers if the measurements have not been contaminated.

I proposed that the pharmacologist use one of these solutions, and he was able to make sense out of the data. Future experiments produced consistent conclusions, showing that the robust analysis was on track.

BOX AND COX

While still at ICI, George Box often visited the statistical group at University College, where he met David Cox. Cox has gone on to become a major innovator in statistics and has been editor of *Biometrika*, Karl Pearson's journal. Both men were struck by the humor of their similar names and by the fact that "Box and Cox" was a term used in English theater to describe two minor parts played by one actor. There is also a classic English musical comedy skit involving two men named Box and Cox who rent the same bed in a rooming house, one of them for the day and the other for the night.

George Box and David Cox resolved to write a paper together. However, their statistical interests were not in the same areas; as the years went by and they tried from time to time to finish the paper, their interests completely diverged, and the paper had to accommodate two different philosophical positions about the nature of statistical analysis. In 1964, the paper was finally published in the *Journal of the Royal Statistical Society*. "Box and Cox," as the paper is known, has since become an important part of statistical methods. In the paper, they show how to convert measurements in a way that increases the robustness of most statistical procedures. "Box-Cox transformations," as they are called, have been used in the analysis of toxicological studies of the mutational effects of chemicals on living cells, in econometric analyses, and even in agricultural research, where R. A. Fisher's methods had originated.

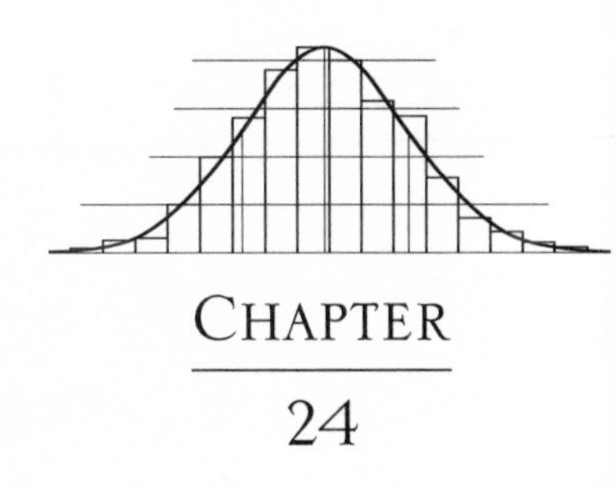

CHAPTER

24

THE MAN WHO REMADE INDUSTRY

In 1980, the NBC television network aired a documentary entitled "If Japan Can, Why Can't We?" The American automobile companies were being rocked by challenges from Japan. The quality of the Japanese automobiles was vastly superior to the American cars of the 1970s, and their prices were lower. Not only in automobiles, but in other areas of industry, from steel to electronics, the Japanese had surpassed American enterprise both in quality and cost. The NBC documentary examined how this had happened. The documentary was really a description of the influence one man had on Japanese industry. That man was an 80-year-old American statistician, W. Edwards Deming.

Suddenly, Deming, who had been working as a consultant to industry since leaving the U.S. Department of Agriculture in 1939, was in great demand. In his long career as a consultant to industry, Deming had been invited many times to help the American automobile companies with quality control. He had developed strong ideas about how to improve industrial methods, but senior management in these companies

W. Edwards Deming, 1900–1993

had no interest in the "technical" details of quality control. They were content to hire experts and have them run quality control—whatever that was. Then, in 1947, General Douglas MacArthur was named the supreme allied commander over conquered Japan. He forced Japan to adopt a democratic constitution and called in the foremost experts on the "American way" to educate the country. His staff located the name of W. Edwards Deming as an expert on statistical sampling methods. Deming was invited to go to Japan to show the Japanese "how we do it in America."

Deming's work impressed Ichiro Ishikawa, the president of the Japanese Union of Scientists and Engineers (JUSE), and he was invited back to teach statistical methods in a group of seminars organized for a broad range of Japanese industry. Ishikawa also had the ear of senior management in many Japanese companies. At his invitation, members of management often attended Deming's lectures. At that time, the phrase "made in Japan" meant cheap, poor-quality imitations of other countries' products. Deming shocked his listeners by telling them they could change that around within five years. He told them that, with the proper use of statistical methods of quality control, they could produce products of such high quality and low price that they would dominate markets all over the world. Deming noted in later talks that he was wrong in predicting this would take five years. The Japanese beat his prediction by almost two years.

The Japanese were so impressed by Deming that JUSE instituted an annual Deming Prize to encourage the development of effective new methods of quality control in Japanese industry. The Japanese government became excited by the prospects of using statistical methods for improvement of all activities, and the education ministry instituted an annual Statistics Day, when students competed for prizes by creating presentations in statistics. Statistical methods swept over Japan—almost all of it derived from the Deming seminars.

DEMING'S MESSAGE TO SENIOR MANAGEMENT

After the 1980 NBC documentary, Deming was welcomed by American industry. He set up a series of seminars to present his ideas to American management. Unfortunately, the senior management of most American companies did not understand what Deming had done, and they sent their technical experts, who already knew about quality control, to his seminars. With very few exceptions, members of the higher levels of management seldom attended. Deming's message was a message to management primarily. It was a nasty, critical message. Management, especially senior management, had been failing to do its job. To illustrate his message, Deming would have the students of his seminar engage in a manufacturing experiment.

The students would be divided into factory workers, inspectors, and managers. The factory workers would be trained in a simple procedure. They would be given a large drum of white beads, with a few red beads mixed in. The training would teach them to mix the drum vigorously by turning it over and over. The training emphasized that the crucial aspect of the job was this mixing. They would then take a paddle with 50 small pits, each big enough to hold one bead. Passing the paddle through the drum, the workers would bring up exactly 50 beads. They were told that marketing had determined that the customers would not tolerate having any more than 3 red beads in the 50, and they were to strive toward that goal. As each worker emerged with a paddle full of beads, the inspector would count the number of red beads and record it. The manager would examine the records and praise those workers whose red-bead counts were small or close to the maximum of 3, and would criticize those workers whose red-bead counts were large. In fact, the manager would often tell the poor workers to stop what they were doing and watch the good workers to see how they were doing it, in order to learn how to do the task correctly.

One-fifth of the beads in the drum were red. The chance of getting 3 or fewer red beads was less than 1 percent. But the chance of getting 6 or fewer was about 10 percent, so the workers would often get tantalizingly close to the magic goal of no more than 3 red beads. On the average, they would get 10 red beads, which was unacceptable to management; and purely by chance some workers would get 13 to 15 red beads, clearly the result of very poor work.

Deming's point was that, all too often, management sets impossible standards and makes no attempt to determine if the standards can be met or what has to be done to change the equipment to meet those standards. Instead, he claimed, American senior management relied on experts in quality control to keep to the standards, ignoring the frustrations the factory workers might have. He was scathingly critical of the fads of management that would sweep through American industry. In the 1970s, the fad was called "zero defect." They would have no defects in their product—a condition that Deming knew was completely impossible. In the 1980s (just as Deming was making an imprint on American industry), the fad was called "total quality management," or TQM. Deming viewed all of this as nothing more than empty words and exhortations from management, who should be doing their real jobs instead.

In his book *Out of the Crisis,* Deming quoted from a report he wrote to the management of one company:

> This report is written at your request after study of some of the problems you are having with low production, high costs, and variable quality. . . . My opening point is that no permanent impact has ever been accomplished in improvement of quality unless the top management carry out their responsibilities. . . . Failure of your own management to accept and act on their responsibilities for quality is, in my opinion, the prime cause of your trouble. . . . What you have in your company . . . is not quality control but guerrilla sniping—no organized system, no provision nor appreciation for control of quality as a system. You have been running a fire department that hopes to arrive in time to keep fires from spreading. . . .
>
> You have a slogan, posted everywhere, urging everyone to do perfect work, nothing else. I wonder how anyone could live up to it. By every man doing his job better? How can he, when he has no way to know what his job is, nor how to do it better? How can he, when he is handicapped with defective materials, change of supply, machines out of order? . . . Another roadblock is management's supposition that the production workers are responsible for all the trouble: that there would be no problems in production if only the production workers would do their jobs in the way that they know to be right.

> . . . in my experience, most problems in production have their origin in common causes, which only management can reduce or remove.

Deming's major point about quality control is that the output of a production line is variable. It has to be variable, because that is the nature of all human activity. What the customer wants, Deming insisted, is not a perfect product but a *reliable* product. The customer wants a product with low variability so he or she can know what to expect from it. R. A. Fisher's analysis of variance allows the analyst to separate the variability of the product into two sources. One source Deming called "special causes." The other he called "common," or "environmental," causes. He claimed that the standard procedure in American industry was to put limits on the total variability allowed. If the variability exceeded those limits, the production line was shut down, and they searched for the special cause of this. But, Deming insisted, the special causes are few and can be easily identified. The environmental causes are always there and are the result of poor management, since they are often in the form of poorly maintained machines, or variable quality in the stock used for manufacturing, or other uncontrolled working conditions.

Deming proposed that the production line be seen as a stream of activities that start with raw material and end with finished product. Each activity can be measured, so each activity has its own variability due to environmental causes. Instead of waiting for the final product to exceed arbitrary limits of variability, the managers should be looking at the variability of each of these activities. The most variable of the activities is the one that should be addressed. Once that variability is reduced, there will be another activity that is "most variable," and *it* should then be addressed. Thus, quality control becomes a continuous process, where the most variable aspect of the production line is constantly being worked on.

The end result of Deming's approach was Japanese cars that ran for 100,000 miles or more without major repairs, ships that needed minimum maintenance, steel that was of consistent quality from batch to batch, and other results of an industry where the variability of quality was under control.

THE NATURE OF QUALITY CONTROL

Walter Shewhart at Bell Telephone Laboratories and Frank Youden at the National Bureau of Standards had brought the statistical revolution into industry by organizing the first programs in statistical quality control in the United States during the 1920s and 1930s. Deming moved the statistical revolution into the offices of upper management. In *Out of the Crisis*, written for managers with minimal mathematical knowledge, he pointed out that many ideas are too vague to use in manufacturing. An automobile piston should be round; but this phrase means nothing unless there is a way to measure the roundness of a particular piston. To improve the quality of a product, the product's quality has to be measured. To measure a property of a product requires that the property (roundness in this case) be well defined. Because all measurements are, by nature, variable, the manufacturing process needs to address the parameters of the distributions of those measurements. Just as Karl Pearson sought to find evidence of evolution in changes in the parameters, Deming insisted that management had the responsibility of monitoring the parameters of these measurement distributions and changing fundamental aspects of the manufacturing process in order to improve those parameters.

I first met Ed Deming at statistical meetings in the 1970s. Tall and stern-looking when he had something critical to say, Deming was a formidable figure among statisticians. Seldom did I see him rise up during the questions after a talk to criticize, but he would often take a person aside after the session to criticize that person for failing to see what, to Deming, was obvious. This stern-faced critic whom I met was not the Deming that his friends knew. I saw his public persona. He was known for his kindness to and consideration for those he worked with, for his robust, if very subtle, humor, and for his interest in music. He sang in a choir, played drums and flute, and published several original pieces of sacred music. One of his musical publications was a new version of the "Star-Spangled Banner"—a version that he claimed was more singable than the usual one.

He was born in Sioux City, Iowa, in 1900, and attended the University of Wyoming, where he studied mathematics with a strong interest in engineering. He received a master's degree in mathematics and physics from the University of Colorado. He met his wife, Agnes Belle, at the university. They moved to Connecticut in 1927, where he began studying for a doctorate in physics at Yale University.

Deming's first industrial job was at the Hawthorne[1] plant of Western Electric in Cicero, Illinois, where he worked during the summers he was at Yale. Walter Shewhart had been laying the groundwork for statistical quality control methods at Bell Labs in New Jersey. Western Electric was part of the same company (AT&T), and attempts were being made to apply Shewhart's methods at the Hawthorne plant, but Deming realized that they had failed to understand Shewhart's message. Quality control was becoming a set of rote manipulations based on setting ranges of variability that were allowable. The ranges were often set so that

[1]The Hawthorne plant gave its name to a phenomenon known as the "Hawthorne effect." An attempt was made to measure the difference between two methods of management during the 1930s at the Hawthorne plant. The attempt failed because the workers improved their efforts immensely for both methods. This was because they knew they were being watched carefully. Since then, the term Hawthorne effect has been used to describe the improvement in a situation that occurs just because an experiment is being run. Typical is the fact that large clinical trials comparing new treatments with traditional ones usually show an improvement in patient health, more than would have been expected from the traditional one based on past experience. This makes it more difficult to detect the difference in effect between the traditional and the new treatment.

5 percent or less of the time a defective product would get through quality control. Deming was later to dismiss this version of quality control as a guarantee that 5 percent of the customers would be dissatisfied.

With his degree from Yale, Deming went to the U.S. Department of Agriculture in 1927, where he worked on sampling techniques and experimental design for the next twelve years. He left government to set up his own consulting company and began running seminars on the use of quality control in industry. These were expanded during World War II, when he trained about 2,000 designers and engineers. These students went on to give seminars at their own companies, until the progeny of Deming-trained quality control specialists reached almost 30,000 by the end of the war.

The last Deming seminar took place on December 10, 1993, in California. Ninety-three-year-old W. Edwards Deming participated, although much of the seminar was run by his younger assistants. On December 20, he died in his home in Washington, D.C. The W. Edwards Deming Institute was founded by his family and friends in November 1993, "to foster understanding of the Deming System of Profound Knowledge to advance commerce, prosperity, and peace."

DEMING ON HYPOTHESIS TESTING

In chapter 11, we saw the development of hypothesis testing by Neyman and Pearson, and how it came to dominate much of modern statistical analysis. Deming was highly critical of hypothesis testing. He ridiculed its widespread use because, he claimed, it focused on the wrong questions. As he pointed out: "The question in practice is never whether a difference between two treatments A and B is significant. Given a difference . . . however small between [them] . . . one can find a . . . number of repetitions of the experiment . . . that will [produce significance]." Thus, to Deming, a finding of a significant difference means nothing. It is the *degree* of difference found that is important. Furthermore, Deming pointed out, the degree of difference found in one experimental situation may not be the same as that found in another setting. To Deming, the standard methods of statistics could not be used, by themselves, to solve problems. These limitations of statistical methods are important. As Deming put it, "Statisticians need to become interested in problems and to learn and teach statistical inference and the limitations thereof. The better we understand the limitations of an inference . . . from a set of results, the more useful becomes the inference."

In the final chapter of this book, we shall look at these limitations of statistical inference that Deming warned about.

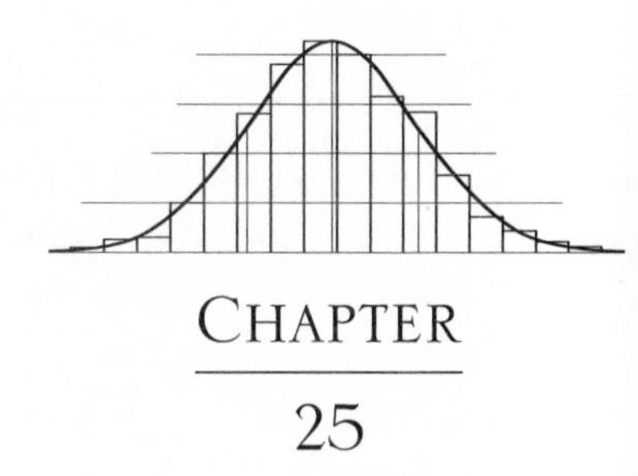

CHAPTER 25

ADVICE FROM THE LADY IN BLACK

Although male theoreticians dominated the development of statistical methods in the early years of the twentieth century, by the time I joined the profession in the 1960s, there were many women in prominent places. This was particularly true in industry and government. Judith Goldberg at American Cyanamid and Paula Norwood at Johnson Pharmaceuticals have headed up statistics departments at pharmaceutical companies. Mavis Carroll was in charge of the mathematical and statistical services division of General Foods. In Washington, D.C., women have been in charge of the Census Bureau, the Bureau of Labor Statistics, and the National Center for Health Statistics, among others. This has also been true in the United Kingdom and continental Europe. In chapter 19, we looked at the roles some of these women have played in the development of statistical methodology.

There is nothing typical about the experiences of women who have made a name for themselves in statistics. All of them are remarkable individuals whose individual developments and accomplishments are unique. One cannot pick a representative woman in statistics, any more than one can find a representative man in statistics. It would be interesting, however, to examine the career of one woman, who rose to prominence in both industry and government. Stella Cunliffe of Great Britain was the first woman to be named president of the Royal Statistical Society. Much of this chapter is taken from the annual address of the president, which she presented before the society on November 12, 1975.

Those who have known or worked with Stella Cunliffe attest to her broad good humor, her keen common sense, and her ability to reduce the most complicated mathematical models to understandable terms for the scientists with whom she collaborated. Much of this comes out in her address. The address is a plea to the members of the Royal Statistical Society to spend less time developing abstract theory and more time collaborating with scientists from other fields. For example, she wrote: "It is no use, as statisticians, our being sniffy about the slapdash methods of many sociologists unless we are prepared to try to guide them into more scientifically acceptable thought. To do this there must be interaction between us." She makes frequent use of examples where the unexpected has occurred in the process of running an experiment. "Barley trials, even on a well-organized research station, could be knocked for six by some fool of a tractor driver hurrying home to his tea via a short cut across the plot."

Stella Cunliffe studied statistics at the London School of Economics in the late 1930s. It was an exciting time to be there. Many of the students and some of the faculty had volunteered to serve in the Spanish Civil War against the fascists. Prominent economists, mathematicians, and other scientists who had escaped Nazi Germany were given temporary positions at the school. When she emerged from school with her degree, the entire world was still suffering from the Great Depression. The only job she could find was with the Danish Bacon Company, "where the use of mathematical statistics was minimal and I, as a statistician, in particular a female statistician, was looked upon as something very odd." With the coming of World War II, Cunliffe became involved in food allocation problems, where her mathematical skills proved very useful.

For two years after the war, she volunteered to help the relief work in devastated Europe. She was one of the first to arrive in Rotterdam, in the Netherlands, while the German army was still surrendering and where the civilian population was starving. She moved on to help with the concentration camp victims at Bergen-Belsen soon after liberation. She finished her efforts working in the displaced-persons camps in the British zone of occupation. Cunliffe returned from her voluntary work penniless and was offered two jobs. One of them was at the Ministry of Food, where she would work in the "oils and fats" department. The other job was at the Guinness Brewing Company, which she accepted. Recall that William Sealy Gosset, who published under the pseudonym of "Student," had founded the statistical department at Guinness. Stella Cunliffe arrived there about ten years after Gosset's death, but his influence was still very strong at Guinness, where his reputation was revered and the experimental discipline he had created dominated its scientific work.

Statistics at Guinness

The Guinness workers believed in their product and the experimentation that was used constantly to improve it. They

> never stopped experimenting to try to produce the product as a constant one, from varying raw materials, varying because of weather, soil, varieties of hops and barley, as economically as possible. They were arrogant about their product and, as may or may not be known, there was no advertising until 1929 because of the attitude—still endemic when I left—that Guinness is the best beer available, it does not need advertising as its quality will sell it, and those who do not drink it are to be sympathized with rather than advertised to!

Cunliffe described her first days at Guinness:

> On arrival at the Dublin Brewery for "training," having led, as I had in Germany, a free and in many ways exciting life, I appeared one morning before the supervisor of the "Ladies Staff" at the Dublin Brewery. She was a forbidding-looking lady, all in black, with small pieces of lace at her throat, held up by whalebones. . . . She impressed on me what a privilege it was to have been chosen to work for Guinness, and reminded me that I was expected to wear stockings and a hat and, if I was lucky enough to meet one of the chosen race known as "brewers" in the corridor, on no account was I to recognize him, but should lower my eyes until he had passed.

Thus the position of women in the hierarchal world of the Guinness Brewing Company in 1946.

Cunliffe soon proved her worth at Guinness and became deeply involved in the agricultural experiments in Ireland. She was not content to stay at her desk and analyze the data sent to her by the field scientists. She went out to the field, to see for herself what was happening. (Any new statistician would do well to follow her example. It is amazing how often the description of an experiment as relayed by somebody several layers above the laboratory workers does not agree with what has actually happened.)

> Many is the damp and chill morning that found me at 7 A.M., shivering and hungry in a hop garden, actually taking part in some vital experiment. I use the word "vital" deliberately because unless the experiment is accepted as vital by the statistician, so that the enthusiasm of the experimenter is shared by the statistician, I submit that his contribution to the work

> is less than optimum. One of the main problems as statisticians is that we have to be flexible: We have to be prepared to switch from helping a microbiologist in the production of a new strain of yeast; to helping an agriculturalist to assess the dung-producing qualities resulting from the intake of particular cattle feeds; to discussing with a virologist the production of antibodies to Newcastle disease; to helping a medical officer assess the effects on health of dust in malt stores; to advising an engineer about his experiments involving a mass-produced article moving along a conveyor belt; to applying queuing theory to the canteen; or to helping a sociologist test his theories about group behavior.

This list of types of collaboration is typical of the work of a statistician in industry. In my own experience, I have had interactions with chemists, pharmacologists, toxicologists, economists, clinicians, and management (for whom we developed operations research models for decision making). This is one of the things that make the work day of a statistician fascinating. The methods of mathematical statistics are ubiquitous, and the statistician, as the expert in mathematical modeling, is able to collaborate in almost every area of activity.

Unexpected Variability

In her address, Stella Cunliffe muses on that greatest source of variability—*Homo sapiens*:

> It was my delight to be responsible for much of the tasting and drinking experiments that are an obvious part of the development of that delicious liquid—Guinness. It was in connection with this that I began to realize how impossible it is to find human beings

> without biases, without prejudices, and without the delightful idiosyncrasies which make them so fascinating. . . . We all have prejudices about certain numbers, letters, or colours, and all of us are very superstitious. We all behave irrationally. I well remember an expensive experiment set up to discover the temperature at which beer was preferred. This involved subjects sitting in rooms at various temperatures, drinking beers at various other temperatures. Little men in white coats ran up and downstairs with beer in buckets of water at varying temperatures, thermometers abounded and an air of bustle prevailed. The beers were identified by coloured crown seals, and the only clear-cut result of this experiment . . . was that our drinking panel showed that the only thing that mattered to them was the colour of the crown seals and that they did not like beer with yellow crown seals.

She describes an analysis of capacity of small beer casks. The casks were handmade and their capacity was measured to determine if they were of appropriate size. The woman who measured them had to weigh the empty cask, fill it with water, and weigh the full cask. If the cask differed from its proper size by being more than three pints below or more than seven pints above, it was returned for modification. As part of the ongoing process of quality control, the statisticians kept track of the filling size of the casks and which ones were discarded. On examining the graph of filling sizes, Cunliffe realized that there was an unusually high number of casks that were just barely making it, and an unusually low number that were just outside the limits. They examined the working conditions of the woman who weighed the casks. She was required to throw a discarded cask onto a high pile and place an accepted cask onto a conveyor belt. At Cunliffe's suggestion, her weighing position was put on top of the bin for the discarded casks. Then all she had to do was kick the rejected cask down into the bin. The excess of casks just barely making it disappeared.

Stella Cunliffe rose to head the statistical department at Guinness. In 1970, she was hired away by the Research Unit of the British Home Office, which supervises police forces, criminal courts, and prisons.

> This unit, when I joined it, was concerned mainly with criminological problems and I plunged . . . straight from the rather precise, carefully designed, thoroughly analyzable work that I had been doing at Guinness into what I can only describe as the airy-fairy world of the sociologist and, if I dare say it, sometimes of the psychologist. . . . I am in no way decrying the ability of the researchers in the Home Office Research Unit. . . . However, it came as a shock to me that those principles of setting up a null hypothesis, of careful experimental design, of adequate sampling, of meticulous statistical analysis, and of detailed assessment of results, with which I had worked for so long, appeared to be much less rigorously applied or accepted in sociological fields.

A great deal of the "research" in criminology was run by accumulating data over time and examining it for the possible effects of public policy. One of these analyses had compared the length of sentence given to adult male prisoners versus the percentage of those men who were reconvicted within two years of discharge. The results clearly showed that the prisoners with short sentences had a much higher rate of recidivism. This was taken as proof that long sentences took "habitual" criminals off the streets.

Cunliffe was not satisfied with a simple table of rates of recidivism versus sentence length. She wanted to look carefully at the raw data behind that table. The strong apparent relationship was due, in large part, to the high recidivism rate among prisoners given sentences of three months or less. But, upon careful examination, almost all these prisoners were "the many old, pathetic, sad and mad people [who] end up in our prisons because the mental hospitals will not take them. They represent a brigade that goes round and round." In fact, because of their frequent incarceration, the same people kept showing up again and again but were being counted as different prisoners when the table was constructed. The rest of the apparent effect of long sentences on recidivism occurred at the other end of the table, where prisoners with sentences of ten years or more had less than a 15 percent rate. "There is a big age factor in this too," she wrote, "a big environmental factor and a big offense one. Large frauds and forgeries tend to attract long sentences—but somebody who has committed a major fraud seldom commits another." Thus, upon her adjusting the table for the two extreme anomalies, the apparent relationship between sentence length and recidivism disappeared.

As she wrote:

> I opine that even the so-called "dull old Home-Office statistics" are fascinating. . . . It seems to me that one of the statistician's jobs is to look at figures, to query why they look like they do. . . . I am being very simple-minded tonight, but I think it is our job to suggest that figures are interesting—and, if the person to whom we say this looks bored, then we have either put it across badly or the figures are not interesting. I suggest that my statistics in the Home Office are not boring.

She decried the tendency for government officials to make decisions without careful examination of the data available:

> I do not think this is the fault of the sociologist, the social engineer, the planner . . . but it must often be laid very

> firmly to the door of the statistician. We have not learned to serve those disciplines that are less scientific than we might like, and we have not therefore been accepted as people who can help them to further knowledge. . . . The strength of the statistician in applied fields, as experienced by me . . . lay in his or her ability to persuade other people to formulate questions which needed answering; to consider whether these questions could be answered with the tools available to the experimenter; to help him set up suitable null hypotheses; to apply rigid disciplines of design to the experiments.

In my own experience, the attempt to formulate a problem in terms of a mathematical model forces the scientist into understanding what question is really being posed. A careful examination of resources available often produces the conclusion that it is not possible to answer that question with those resources. I think that some of my major contributions as a statistician were when I discouraged others from attempting an experiment that was doomed to failure for lack of adequate resources. For instance, in clinical research, when the medical question posed will require a study involving hundreds of thousands of patients, it is time to reconsider whether that question is worth answering.

ABSTRACT MATHEMATICS VERSUS USEFUL STATISTICS

Stella Cunliffe emphasized the hard work of making statistical analyses useful. She was disdainful of elaborate mathematics for the sake of mathematics and decried mathematical models that are

> all the imagination and lack of reality . . . lots of string, interesting side-pieces, plenty of amusement, brilliant of

> concept, but also the same lack of robustness and of reality. The delight in elegance, often at the expense of practicality, appears to me, if I dare say so, to be rather a male attribute. . . . We statisticians are educated to calculate . . . with mathematical precision . . . [but] we are not good at . . . persuading the uninitiated that our findings are worth heeding. We shall not succeed in so doing if we solemnly quote "p less than 0.001" to an incomprehending man or woman; we must explain our findings in their language and develop the powers of persuasion.

Without a hat, and refusing to make herself meekly subordinate to the master brewers, Stella Cunliffe flew into the world of statistics, jauntily indulging her lively curiosity and criticizing the professors of mathematical statistics who came to hear her speech. As of this writing, she could still be found attending meetings of the Royal Statistical Society and skewering mathematical pretensions with her tart wit.

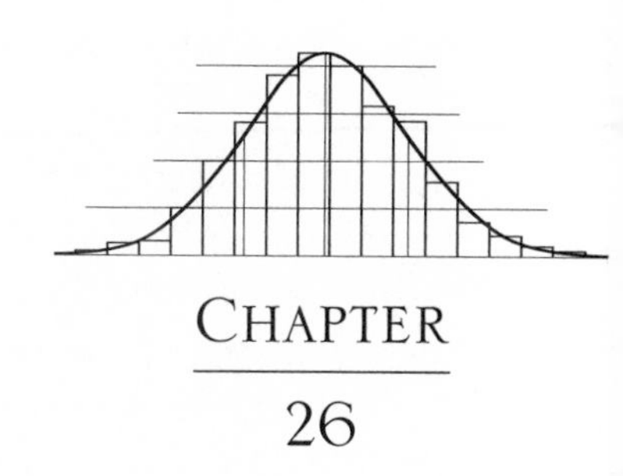

Chapter 26

The March of the Martingales

Congestive heart failure is one of the major causes of death in the world. Although it often attacks men and women in the prime of life, it is primarily a disease of old age. Among citizens of the United States above the age of sixty-five, congestive heart failure or its complications account for almost half the deaths. From the standpoint of public health, congestive heart failure is more than a cause of death; it is also a cause of considerable illness among the living. The frequent hospitalizations and the complicated medical procedures used to stabilize patients with congestive heart failure are a major factor in the overall cost of medical services in the country. There is an intense interest in finding effective outpatient care that can reduce the need for hospitalizations and improve the quality of life for these patients.

Unfortunately, congestive heart failure is not a simple disease that can be attributed to a single infective agent or which can be alleviated by blocking a particular enzymatic pathway. The primary symptom of congestive heart failure is the increasing weakness of the heart muscle. The heart becomes less and less able to respond to the subtle commands of hormones that regulate its rate and strength of contraction to meet the changing needs of the body. The heart muscle becomes enlarged and flabby. Fluid builds up in the lungs and ankles. The patient becomes breathless after slight exertion. The reduced amount of blood being pumped through the body means that the brain has a reduced level when the stomach demands blood to digest a meal, and the patient becomes confused or dozes for long periods of time.

To maintain homeostasis, the life forces in the patient adapt to this decrease in heart output. In many patients, the balance of hormones that regulate the heart and other muscles changes to reach a somewhat stable state, where some of the hormonal levels and their responses are "abnormal." If the physician treats this abnormal balance with drugs like beta-adrenergic agonists or calcium channel blockers, the result may be an improvement in the patient's condition. Or, by tipping over that barely stable state, the treatment may drive the patient into further deterioration. One of the major causes of death among congestive heart failure patients used to be the buildup of fluid in their lungs (formerly called dropsy). Modern medicine makes use of powerful diuretics, which keep the fluid level down. In the process, however, these diuretics can, themselves, introduce problems in the feedback between hormones generated by the kidney and those generated by the heart in response.

The search continues for effective medical treatments to prolong the life of these patients, reduce the frequency of hospitalizations, and improve the quality of their lives. Since some treatments may have counterproductive effects on some patients, any clinical study of these treatments will have to take specific patient characteristics into account. In this way, the final analysis of data from such a study can identify those patients for whom the treatment is effective and those patients for whom it is countereffective. The statistical analysis of congestive heart failure studies can become exceedingly difficult.

When designing such a study, the first question is what to measure. We could, for instance, measure the average number of hospitalizations of patients on a given treatment. This is a rough overall measure that misses such important aspects as the patients' ages, their initial health states, and the frequency and length of those hospitalizations. It would be better to consider the time course of each patient's disease, accounting for the hospitalizations that might occur, how long they last, how long since the previous one, measurements of quality of life between hospitalizations, and adjusting all these outcomes for the patient's age and other diseases that might be present. This might be the ideal from a medical point of view, but it poses difficult statistical problems. There is no single number to associate with each patient. Instead, the patient's record is a time course of events, some of them repeated, others of which are measured by multiple measurements. The "measurements" of this experiment are multileveled, and the distribution function, whose parameters must be estimated, will have a multidimensional structure.

Early Theoretical Work

The solution to this problem begins with the French mathematician Paul Lévy. Lévy was the son and grandson of mathematicians. Born in 1886, he was identified early on as a gifted student. Following the usual procedure in France at the time, he was quickly moved through a series of special schools for the gifted and won academic honors. He received the Prix du Concours Général in Greek and mathematics while still in his teens; the Prix d'Excellence in mathematics, physics, and chemistry at Lycée Saint Louis; and a first Concours d'Entrée at the Ecole Normale Supérieure and at the Ecole Polytechnique. In 1912, at age twenty-six, he received his docteur des sciences degree, and his thesis was the basis of a major book he wrote on abstract functional analysis. By the time he was thirty-three, Paul Lévy was a full professor at the

Ecole Polytechnique and a member of the Académie des Sciences. His work in the abstract theories of analysis made him world famous. In 1919, he was asked by his school to prepare a series of lectures on probability theory, and he began examining that subject in depth for the first time.

Paul Lévy was dissatisfied with probability theory as a collection of sophisticated counting methods. (Andrei Kolmogorov had not yet made his contribution.) Lévy looked for some underlying abstract mathematical concepts that might allow him to unify these many methods. He was struck by de Moivre's derivation of the normal distribution and by the "folk theorem" among mathematicians that de Moivre's result should hold for many other situations—what came to be called the "central limit theorem." We have seen how Lévy (along with Lindeberg in Finland) in the early 1930s finally proved the central limit theorem and determined the conditions necessary for it to hold. In doing so, Lévy started with the formula for the normal distribution and worked backward, asking what were the unique properties of this distribution that would make it rise out of so many situations.

Lévy then approached the problem from the other direction, asking what was it about specific situations that led to the normal distribution. He determined that a simple set of two conditions will guarantee that data tend to be normally distributed. These two conditions are not the only ways the normal distribution can be generated, but Lévy's proof of the central limit theorem established the more general set of conditions that is always needed. These two conditions were adequate for the situation where we have a sequence of randomly generated numbers, one following on the other:

1. The variability has to be bounded so individual values do not become infinitely large or small.

2. The best estimate of the next number is the value of the last number.

Lévy called such a sequence a "martingale."

Lévy appropriated the word *martingale* from a gambling term. In gambling, a martingale is a procedure wherein the gambler doubles his bet each time he loses. If he has a 50:50 chance of winning, the expected loss is equal to his previous loss. There are two other meanings to the word. One describes a device used by French farmers to keep a horse's head down and to keep the animal from rearing. The farmer's martingale keeps the horse's head in a position so that it can be moved at random, but the most probable future position is the one the head is held in now. A third definition of the term is a nautical one. A martingale is a heavy piece of wood hung from the jib of a sail to keep the jib from swinging too far from one side to the other. Here, too, the last position of the jib is the best predictor of its next position. The word itself is derived from the inhabitants of the French town of Martique, who were legendary for their stinginess, so the best estimate of the little money they would give next week was the little they gave today.

Thus, the stingy inhabitants of Martique gave their name to a mathematical abstraction in which Paul Lévy developed the stingiest possible characteristics of a sequence of numbers that tend to have a normal distribution. By 1940, the martingale had become an important tool in abstract mathematical theory. Its simple requirements meant that many types of sequences of random numbers could be shown to be martingales. In the 1970s, Odd Aalen of the University of Oslo in Norway realized that the course of patient responses in a clinical trial was a martingale.

Martingales in Congestive Heart Failure Studies

Recall the problems arising from a congestive heart failure study. The patient responses tend to be idiosyncratic. There are questions about how to interpret events like hospitalizations when they occur early in the study or later (when the patients have become older).

There are questions about how to deal with the frequency of hospitalizations and the length of stay in the hospitals. All of these questions can be answered by viewing as martingales the stream of numbers taken over time. In particular, Aalen noted, a patient who is hospitalized can be taken out of the analysis and returned to it when released. Multiple hospitalizations can be treated as if each one were a new event. At each point in time, the analyst need know only the number of patients still in the study (or returned to the study) and the number of patients who were initially entered into the study.

By the early 1980s, Aalen was working with Erik Anderson of the University of Aarhus in Denmark and Richard Gill of the University of Utrecht in the Netherlands, exploiting the insight that he had developed. In the first chapter of this book, I pointed out that scientific and mathematical research is seldom done alone. The abstractions of mathematical statistics are so involved that it is easy to make mistakes. Only by discussion and criticism among colleagues can many of these mistakes be found. The collaboration among these three, Aalen, Anderson, and Gill, provided one of the most fruitful developments of the subject in the final decades of the twentieth century.

The work of Aalen, Anderson, and Gill has been supplemented by that of Richard Olshen and his collaborators at the University of Washington and by Lee-Jen Wei at Harvard University, to produce a wealth of new methods for analyzing the sequence of events that occur in a clinical trial. L. J. Wei, in particular, has exploited the fact that the difference between two martingales is also a martingale, to eliminate the need to estimate many of the parameters of the model. Today, the martingale approach dominates the statistical analysis of long-term clinical trials of chronic disease.

The legendary stinginess of the inhabitants of Martique was the starting point. A Frenchman, Paul Lévy, had the initial insights. The mathematical martingale passed through many other minds, with contributions from Americans, Russians, Germans, English, Italians, and Indians. A Norwegian, a Dane, and a Dutchman brought it to clinical research. Two Americans, one born in Taiwan, elaborated on their work. A complete listing of the authors of papers and books on this topic, which have emerged since the late 1980s, would fill many pages and involve workers from still other countries. Truly, mathematical statistics has become an international work in the making.

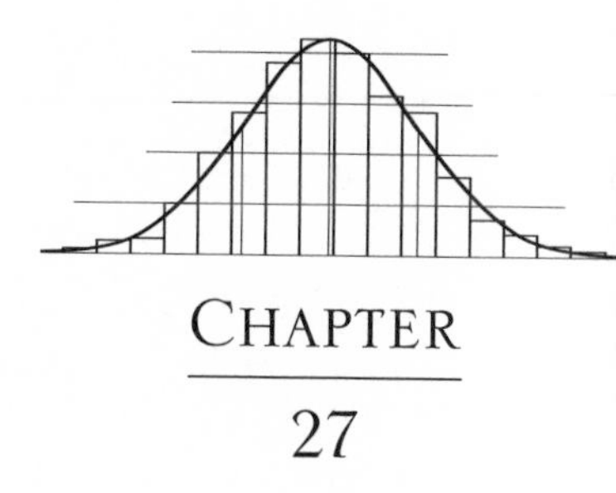

CHAPTER

27

THE INTENT TO TREAT

In the early 1980s, Richard Peto, one of the leading biostatisticians in Great Britain, had a problem. He was analyzing the results of several clinical trials comparing different treatments for cancer among patients. Following the dictates of R. A. Fisher's experimental design, the typical clinical trial identified a group of patients who were in need of treatment and assigned them, at random, to different experimental methods of treatment.

The analysis of such data should have been relatively straightforward. The percentage of patients who survived for five years would have been compared among the treatment groups, using Fisher's methods. A more subtle comparison could have been made, using Aalen's martingale approach to analyze the time from the beginning of the study to each patient's death as the basic measure of effect. Either way, the analysis was based on the initial randomization of patients to treatment. Following Fisher's dictums, the assignment of patients to treatment was completely independent of the outcome of the study, and p-values for hypothesis tests could be calculated.

Peto's problem was that all the patients had not been given the treatments to which they were randomized. These were human beings, suffering from painful and, in many cases, terminal disease. The physicians treating them felt impelled to abandon the experimental treatment or at least to modify it if they felt it was in the best interests of the patient. Blind following of an arbitrary treatment, without considering the patient's needs and responses, would have been unethical. Contrary to Fisher's dictums, the patients in these studies were often provided with new treatments wherein the choice of treatment *did* depend upon the patient's response.

This was a typical problem in cancer studies. It had been a problem in such studies since they were first started in the 1950s. Until Peto came on the scene, the usual procedure was to analyze only those patients who remained on their randomized treatment. All other patients were dropped from the analysis. Peto realized that this could lead to serious errors. For instance, suppose we were comparing an active treatment with treatment by a placebo, a drug that has no biological effect. Suppose that patients who fail to respond are switched to a standard treatment. The patients who fail on placebo will be switched over and left out of the analysis. The only patients who remain on placebo will be patients who, for some other reason, are responding. The placebo will be found to be as effective (or perhaps even more so) than the active treatment—if the patients who remained on placebo and responded are the only placebo-treated patients used in the analysis.

Edmund Gehan, who was at M. C. Anderson Hospital in Texas, had seen the problem before Peto. His solution at the time was to propose that these studies did not meet Fisher's requirements, so they could not be considered useful experiments for comparing treatments. Instead, the records from these studies consisted of careful observations taken on patients given different types of treatments. The best that could be expected would be a general description of their outcome, with hints of possible future treatment. Later, Gehan considered various other solutions to this problem, but his first conclusion reflects the frustration of someone who tries to apply the methods of statistical analysis to a poorly designed or executed experiment.

Peto suggested a straightforward solution. The patients had been randomized to receive specific treatments. The act of randomization is what made it possible to calculate the p-values of hypothesis tests comparing those treatments. He suggested that each patient be treated in the analysis as if he or she had been on the treatment to which he or she had been randomized. The analyst would ignore all treatment changes that occurred during the course of the study. If a patient was randomized to treatment A and was moved off of that treatment just before the end of the study, that patient was analyzed as a treatment A patient. If the patient randomized to treatment A was on treatment A for only a week, that patient was analyzed as a treatment A patient. If the patient randomized to treatment A never took a single pill from treatment A but was hospitalized and put on alternative therapies immediately after entering the study, that patient was analyzed as a treatment A patient.

This approach may seem foolish at first glance. One can produce scenarios in which a standard treatment is being compared to an experimental one, with patients switched to the standard if they fail. Then, if the experimental treatment is worthless, all or most of the patients randomized to it will be switched to the standard, and the analysis will find the two treatments the same. As Richard Peto made it clear in his proposal, this method of analyzing the results of a study cannot be used to find that treatments are equivalent. It can only be used if the analysis finds that they *differ* in effect.

Peto's solution came to be called the "intent to treat" method. The justification for this name and for its use in general was the following: If we are interested in the overall results of a medical policy that would recommend the use of a given treatment, the physician has to be given the freedom to modify treatment as she sees fit. An analysis of a clinical trial, using Peto's solution, would determine if it is good public policy to recommend a given treatment as a starting treatment. The application of the intent to treat method of analysis was proposed as a sensible one for large government-sponsored studies designed to determine good public policies.

Unfortunately, there is a tendency for some scientists to use statistical methods without knowing or understanding the mathematics behind them. This often appears in the world of clinical research. Peto had pointed out the limitations of his solution. In spite of this, the intent to treat method became enshrined in medical doctrine at many universities and came to be seen as the only correct method of statistical analysis of a clinical trial. Many clinical trials, especially those in cancer, are designed to show that a new treatment is at least as good as the standard, while presenting fewer side effects. The purpose of many trials is to show therapeutic equivalence. As Peto pointed out, his solution can be used only to find differences, and failure to find differences does not mean that the treatments are equivalent.

The problem lay, to some extent, in the rigidity of the Neyman-Pearson formulation. The standard redaction of the Neyman-Pearson formulation found in elementary statistics textbooks tends to present hypothesis testing as a cut-and-dried procedure. Many purely arbitrary aspects of the methods are presented as immutable.

While many of these arbitrary elements may not be appropriate for clinical research,[1] the need that some medical scientists have to

[1]In 1963, Francis Anscombe from Yale University proposed an entirely different approach that would be more in keeping with medical needs. The Neyman-Pearson formulation preserves the proportion of times the analyst will be wrong. Anscombe asked why the statistical analyst's long-run probability of error had anything to do with deciding whether a medical treatment is effective. Instead, Anscombe proposed that there are a finite number of patients who will be treated. A small number of them will be treated in a clinical trial. The rest will be given the treatment that the clinical trial decides is "best." If we use too small a number of patients in the trial, the decision of which treatment is best can be in error, and, if so, all the rest of the patients will be given the wrong treatment. If we use too many patients in the trial, then all the trial patients on the other treatments (not the "best" one) will have been put on the wrong treatment. Anscombe proposed that the criteria of analysis should be minimizing the total number of patients (both those in the trial and those who are treated afterward) who are given the poorer treatment.

use "correct" methods has enshrined an extremely rigid version of the Neyman-Pearson formulation. Nothing is acceptable unless the p-value cutoff is fixed in advance and preserved by the statistical procedure. This was one reason why Fisher opposed the Neyman-Pearson formulation. He did not think that the use of p-values and significance tests should be subjected to such rigorous requirements. He objected, in particular, to the fact that Neyman would fix the probability of a false positive in advance and act only if the p-value is less than that. Fisher suggested in his book *Statistical Methods and Scientific Inference* that the final decision about what p-value would be significant should depend upon the circumstances. I used the word *suggested*, because Fisher is never quite clear on how he would use p-values. He only presents examples.

Cox's Formulation

In 1977, David R. Cox (of Box and Cox from chapter 23) took up Fisher's arguments and extended them. To distinguish between Fisher's use of p-values and the Neyman-Pearson formulation, he called Fisher's method "significance testing" and the Neyman-Pearson formulation "hypothesis testing." By the time Cox wrote his paper, the calculation of statistical significance (through the use of p-values) had become one of the most widely used methods of scientific research. Thus, Cox reasoned, the method has proven to be useful in science. In spite of the acrimonious dispute between Fisher and Neyman, in spite of the insistence of statisticians like W. Edwards Deming that hypothesis tests were useless, in spite of the rise of Bayesian statistics, which had no place for p-values and significances—in spite of all these criticisms among mathematical statisticians, significance testing and p-values are constantly being used. How, Cox asked, do scientists actually use these tests? How do they know that the results of such tests are true or useful? He discovered that, in practice, scientists use hypothesis tests primarily for refining their views of reality by eliminating unnecessary parameters, or for deciding between two differing models of reality.

BOX'S APPROACH

George Box (the other half of Box and Cox) approached the problem from a slightly different perspective. Scientific research, he noted, consisted of more than a single experiment. The scientist arrives at the experiment with a large body of prior knowledge or at least with a prior expectation of what might be the result. The study is designed to refine that knowledge, and the design depends upon what type of refinement is sought. Up to this point, Box and Cox are saying much the same thing. To Box, this one experiment is part of a stream of experiments. The data from this experiment are compared to data from other experiments. The previous knowledge is then reconsidered in terms of both the new experiment and new analyses of the old experiments. The scientists never cease to return to older studies to refine their interpretation of them in terms of the newer studies.

As an example of Box's approach, consider the manufacturer of paper who is using one of Box's major innovations, evolutionary variation in operations (EVOP). With Box's EVOP, the manufacturer introduces experiments into the production run. The humidity, speed, sulfur, and temperature are modified slightly in various ways. The resulting change in paper strength is not great. It cannot be great and still produce a salable product. Yet these slight differences, subjected to Fisher's analysis of variance, can be used to propose another experiment, in which the average strength across all the runs has been slightly increased, and the new runs are used to find the direction of still another slight increase in strength. The results of each stage in EVOP are compared to previous stages. Experiments that seem to produce anomalous results are rerun. The procedure continues forever—there is no final "correct" solution. In Box's model, the sequence of scientific experiments followed by examination and reexamination of data has no end—there is no final scientific truth.

Deming's View

Deming and many other statisticians have rejected the use of hypothesis tests outright. They insist that Fisher's work on methods of estimation should form the basis of statistical analyses. It is the parameters of the distribution that should be estimated. It makes no sense to run analyses that deal indirectly with these parameters through p-values and arbitrary hypotheses. These statisticians continue to use Neyman's confidence intervals to measure the uncertainty of their conclusions; but Neyman-Pearson hypothesis testing, they contend, belongs in the waste bin of history, along with Karl Pearson's method of moments. It is interesting to note that Neyman, himself, seldom used p-values and hypothesis tests in his own applied papers.

This rejection of hypothesis testing, and the Box and Cox reformulations of Fisher's concept of significance testing may cast doubt on Richard Peto's solution to the problem he found in clinical cancer studies. But the basic problem he faced remains. What do you do when the experiment has been modified by allowing the consequences of the treatment to change the treatment? Abraham Wald had shown how a particular type of modification can be accommodated, leading to sequential analysis. In Peto's case, the oncologists were not following Wald's sequential methods. They were inserting different treatments as they perceived the need.

Cochran's Observational Studies

In some ways, this is a problem that William Cochran of Johns Hopkins University dealt with in the 1960s. The city of Baltimore wanted to determine if public housing had an effect on the social attitudes and progress of poor people. They approached the statistics group at Johns Hopkins to help them set up an experiment. Following Fisher's methods, the Johns Hopkins statisticians suggested they take a group of people, whether they had applied for public housing or not, and randomly assign some of them to public housing and refuse it to the others. This horrified the city officials. When openings were announced in public housing, it was their practice to respond on a first-come first-served basis. It was only fair. They could not deny their rights to people who rushed to be "first"—and that on the basis of a computer-generated randomization. The Johns Hopkins statistics group pointed out, however, that those who rushed to apply were often the most energetic and ambitious. If this was true, then those in public housing would do better than the others—without the housing itself having any effect.

Cochran's solution was to propose that they were not going to be able to use a designed scientific experiment. Instead, by following families who went into public housing and those who did not, they would have an observational study, where the families differed by many factors, such as age, educational level, religion, and family stability. He proposed methods for running a statistical analysis of such observational studies. He would do this by adjusting the outcome measurement for a given family to take these different factors into account. He would set up a mathematical model in which there would be an effect due to age, an effect due to whether it was an intact family, an effect due to religion, and so forth. Once the parameters of all these effects had been estimated, the remaining differences in effect would be used to determine the effect of public housing.

When a clinical study announces that the difference in effect has been adjusted for patient age or patient sex, this means that researchers have applied some of Cochran's methods to estimate the underlying effect of treatment, taking into account the effect of imbalances in the assignment of treatment to patients. Almost all sociological studies use Cochran's methods. The authors of these studies may not recognize them as coming from William Cochran, and many of the specific techniques often predate his work. Cochran put it on a solid theoretical foundation, and his papers on observational studies have influenced medicine, sociology, political science, and astronomy—all areas in which random assignment of "treatment" is either impossible or unethical.

RUBIN'S MODELS

In the 1980s and 1990s, Donald Rubin of Harvard University proposed a different approach to Peto's problem. In Rubin's model, each patient is assumed to have a possible response to each of the treatments. If there are two treatments, each patient has a potential response to both treatment A and treatment B. We can observe the patient under only one of these treatments, the one to which the patient had been assigned. We can set up a mathematical model in which there is a symbol in the formula for each of those possible responses. Rubin derived conditions on this mathematical model that are needed to estimate what might have happened had the patient been put on the other treatment.

Rubin's models and Cochran's methods can be applied in modern statistical analyses because they make use of the computer to engage in massive amounts of number crunching. Even if they had been proposed during Fisher's time, they would not have been feasible. They require the use of the computer because the mathematical models are highly involved and complicated. They often require iterative techniques, where the computer does thousands or even millions of estimations, the sequence of estimations converging on the final answer.

These Cochran and Rubin methods are highly model specific. That is, they will not produce correct answers unless the complicated mathematical models they use come close to describing reality. They require the analyst to devise a mathematical model that will match reality in all or most of its aspects. If the reality does not match the model, then the results of the analysis may not hold. A concomitant part of approaches like those of Cochran and Rubin has been the effort to determine the degree to which the conclusions are robust. Current mathematical investigations are looking at how far reality can be from the model before the conclusions are no longer true. Before he died in 1980, William Cochran was examining these questions.

Methods of statistical analysis can be thought of as lying on a continuum, with highly model-bound methods like those proposed by Cochran and Rubin on one end. At the other end, there are nonparametric methods, which examine data in terms of the most general type of patterns. Just as the computer has made highly model-bound methods feasible, there has been a computer revolution at the other end of statistical modeling—this nonparametric end where little or no mathematical structure is assumed and the data are allowed to tell their story without forcing them into preconceived models. These methods go by fanciful names like "the bootstrap." They are the subject of the next chapter.

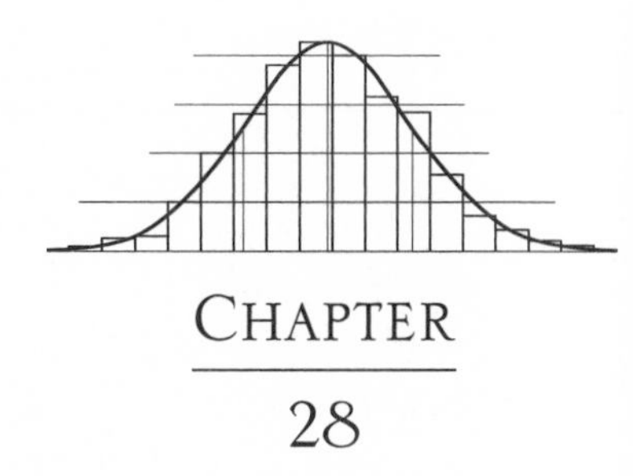

CHAPTER

28

THE COMPUTER TURNS UPON ITSELF

Guido Castelnuovo came from a proud Italian Jewish family. They could trace their ancestry back to ancient Rome during the time of the first Caesars. As a member of the faculty of mathematics at the University of Rome in 1915, Castelnuovo was fighting a lonely battle. He wanted to introduce courses in probability and the mathematics of actuarial studies into the graduate program. At that time, before Andrei Kolmogorov had laid the foundations of probability theory, mathematicians viewed probability as a collection of methods that made use of complicated counting techniques. It was an interesting sidelight of mathematics, often taught as part of an algebra course, but hardly worth consideration in a graduate program at a time when the beautiful shimmering abstractions of pure mathematics were being codified. As far as actuarial mathematics were concerned, this was applied mathematics at its worst, calculations of life spans and accident frequencies computed with relatively simple arithmetic. So the other members of the faculty believed.

In addition to his pioneering work in the abstract field of algebraic geometry, Castelnuovo had a strong interest in applications of mathematics, and he persuaded the rest of the faculty to allow him to construct such a course. As a result of teaching this course, Castelnuovo published one of the first textbooks on probability with statistical applications, *Calcolo della probabilità e applicazioni*, published in 1919. This book was used in similar courses that were presented at other universities in Italy. By 1927, Castelnuovo had founded the School of Statistics and Actuarial Sciences at the University of Rome, and throughout the 1920s and 1930s there were lively exchanges between the growing Italian school of statisticians involved in actuarial research and a similar group in Sweden.

In 1922, Benito Mussolini had brought fascism to Italy and imposed rigid controls on freedom of expression. Both students and faculty at universities were examined to exclude "enemies of the state." There were no racial components to those exclusions, and the fact that Castelnuovo was Jewish was not under consideration.[1] He was able to continue his work for the first eleven years of the fascist government. In 1935, the pact between the Italian fascists and the German Nazis led to the imposition of anti-Semitic laws in Italy, and seventy-year-old Guido Castelnuovo was dismissed from his post.

That did not end the career of this tireless man, who lived until 1952. With the coming of Nazi-inspired racial laws, many promising Jewish graduate students were also dismissed from the universities. Castelnuovo organized special courses in his home, and in the homes of other Jewish former professors, to enable the graduate students to continue their studies. In addition to writing books on the history of mathematics, Castelnuovo spent the last

[1]In its initial form, Italian fascism was strongly profamily. Because of this, only married men were allowed to hold posts in government. These included positions on university faculties. In 1939, brilliant Bruno de Finetti won a nationwide competition for a position as full professor of mathematics at the University of Trieste, but he was not allowed to take it since he was, at that time, still a bachelor.

of his eighty-seven years examining the philosophical relationships between determinism and chance and trying to interpret the concept of cause and effect—topics that we have touched upon in previous chapters and that I shall examine further in the final chapter of this book.

The Italian school of statistics that emerged from Castelnuovo's efforts had solid mathematical foundations but used problems from actual applications as the starting point of most investigations. Castelnuovo's younger contemporary, Corrado Gini, headed the Istituto Centrale di Statistica in Rome, a private institution organized by the insurance companies to further actuarial research. Gini's lively interest in all kinds of applications brought him in touch with most of the young Italian mathematicians involved in mathematical statistics during the 1930s.

THE GLIVENKO-CANTELLI LEMMA

One of these mathematicians was Francesco Paolo Cantelli (1875–1966), who almost anticipated Kolmogorov in establishing the foundations of probability theory. Cantelli was not that interested in investigating foundation issues (dealing with questions like, What is the meaning of probability?), and he failed to plunge as deeply into the underlying theory as Kolmogorov. Cantelli was content to derive fundamental mathematical theorems based on the type of probability calculations that had been around since Abraham de Moivre introduced the methods of calculus into probability calculations in the eighteenth century. In 1916, Cantelli discovered what has been called the fundamental theorem of mathematical statistics. In spite of its great importance, it goes by the unassuming name of the "Glivenko-Cantelli lemma."[2] Cantelli was the first to

[2]During the eighteenth century, the formal mathematics of Euclid's *Elements* were translated into geometry textbooks, and the patterns of logical deduction were codified. Under this codification, the word *theorem* was used to describe a conclusion that was specific to the problem at hand. In order to prove some theorems, it was often necessary to prove intermediate results that could be used in this final theorem, but would also be available for the proof of other theorems. Such a result was called a lemma.

prove the theorem, and he fully understood its importance. Joseph Glivenko, a student of Kolmogorov, gets partial credit because he made use of a newly developed mathematical notation known as the "Stieltjes integral" to generalize the result in a paper he published (in an Italian mathematical journal) in 1933. Glivenko's notation is the one used most often in modern textbooks.

The Glivenko-Cantelli lemma is one of those results that appears to be intuitively obvious, but only after it was discovered. If nothing is known about the underlying probability distribution that generated a set of data, the data themselves can be used to construct a nonparametric distribution. This is an ugly mathematical function, filled with discontinuities and lacking in any kind of elegance. But in spite of its awkward structure, Cantelli was able to show that this ugly empirical distribution function got closer and closer to the true distribution function as the number of observations increased.

The importance of the Glivenko-Cantelli lemma was recognized immediately, and throughout the next twenty years, many important theorems were proven by reducing them to repeated applications of this lemma. It was one of those tools of mathematical research that can almost always be used in a proof. To use the lemma, mathematicians during the first part of the century had to produce clever manipulations of counting techniques. Construction of an empirical distribution function consists of a sequence of mindless steps in simple arithmetic. Without clever tricks, use of the empirical distribution function in estimating parameters from large samples of data would take a fantastic mechanical computer capable of millions of operations a second. No such machine existed in the 1950s, or in the 1960s, or even in the 1970s. By the 1980s, computers had reached the point where this was feasible. The Glivenko-Cantelli lemma became the basis of a new statistical technique that could only exist in a world of high-speed computers.

EFRON'S "BOOTSTRAP"

In 1982, Bradley Efron of Stanford University invented "the bootstrap." It is based on two simple applications of the Glivenko-Cantelli lemma. The applications are simple conceptually, but they require extensive use of the computer to calculate, recalculate, and recalculate again. A typical bootstrap analysis on a moderately large set of data might take several minutes even with the most powerful computer.

Efron called this procedure the bootstrap because it was a case of the data lifting themselves up by their own bootstraps, so to speak. It works because the computer does not mind doing mindless, repetitive arithmetic. It will do the same thing over and over and over again and never complain. With the modern transistor-based chip, it will do so in a few microseconds. There are some complicated mathematics behind Efron's bootstrap. His original papers proved that this method is equivalent to the standard methods if certain assumptions are made about the true underlying distribution. The implications of this method have been so extensive that almost every issue of the mathematical statistical journals since 1982 has contained one or more articles involving the bootstrap.

RESAMPLING AND OTHER COMPUTER-INTENSIVE METHODS

There are alternative versions of the bootstrap and related methods, all of which go under the general name of resampling. In fact, Efron has shown that many of R. A. Fisher's standard statistical methods can be viewed as forms of resampling. In turn, resampling is part of a wider range of statistical methods, which are called "computer intensive." Computer-intensive methods make use of the ability of the modern computer to engage in vast amounts of calculations, working the same data over and over again.

One such procedure was developed during the 1960s by Joan Rosenblatt at the National Bureau of Standards and by Emmanuel Parzen at Iowa State University, working independently of one another. Their methods are known as "kernel density estimation." Kernel density estimation, in turn, led to kernel density-based regression estimation. These methods include two arbitrary parameters, called the "kernel" and the "bandwidth." Soon after these ideas appeared, in 1967 (long before there were computers powerful enough to take full advantage of them), John van Ryzin of Columbia University used the Glivenko-Cantelli lemma to determine the optimal configuration of these parameters.

While the mathematical statisticians were generating theories and writing in their own journals, Rosenblatt's and Parzen's kernel density-based regression was discovered independently by the engineering community. Among computer engineers, it is called "fuzzy approximation." It uses what van Ryzin would have called a "nonoptimal kernel" and there is only one, very arbitrary, choice of bandwidth. Engineering practice is not based on seeking the theoretically best possible method. Engineering practice is based on what will work. While the theoreticians were worrying about abstract optimality criteria, the engineers were out in the real world, using fuzzy approximations to produce computer-based fuzzy systems. Fuzzy engineering systems are used in smart cameras, which automatically adjust the focus and iris. They are also used in new buildings to maintain constant and comfortable temperatures, which may vary according to different needs in different rooms.

Bart Kosko, a private consultant in engineering, is one of the most prolific popularizers of fuzzy systems. As I look over the bibliographies in his books, I can find references to mainstream mathematicians of the nineteenth century like Gottfried Wilhelm von Leibniz, along with references to the mathematical statistician Norbert Wiener, who contributed to the theory of stochastic processes and their applications to engineering methods. I cannot find references to Rosenblatt, Parzen, van Ryzin, or any of the later contributors to the theory of kernel-based regression. Coming up with almost exactly the same computer algorithms, fuzzy systems and kernel density-based regression appear to have been developed completely independently of one another.

THE TRIUMPH OF STATISTICAL MODELS

This extension of computer-intensive statistical methods to standard engineering practice is an example of how the statistical revolution in science had become so ubiquitous by the end of the twentieth century. The mathematical statisticians are no longer the sole or even the most important participants in its development. Many of the fine theories that have appeared in their journals over the past seventy years are unknown to the scientists and engineers who, even so, make use of them. The most important theorems are often rediscovered, again and again.[3]

Sometimes the basic theorems are not even re-proved, but the users assume them to be true because they seem intuitively true. In a few cases, the users invoke theorems that have been proved to be false—again, because they seem intuitively true. This is because the concepts of probability distributions have become so ingrained in modern scientific education that the scientists and engineers think in terms of distributions. Over a hundred years ago, Karl Pearson proposed that all observations arise from probability distributions and that the purpose of science is to estimate the parameters of those distributions. Before that, the world of science believed that the universe followed laws, like Newton's laws of motion, and that any apparent variations in what was observed were due to errors.

[3]My own Ph.D. thesis made use of a class of distributions known, at least among statisticians, as "compound-Poisson distributions." As I worked on my thesis, I had to survey the literature, and I found the same distribution in economics, operations research, electrical engineering, and sociology. In some places it was called the "stuttering Poisson." In other places it was the "Poisson-binomial." In one paper, it was the "Fifth Avenue bus distribution."

Gradually, Pearson's view has become the predominant one. As a result, anyone trained in scientific methods in the twentieth century takes the Pearsonian view for granted. It is so ingrained in modern scientific methods of data analysis that few even bother to articulate it. Many working scientists and engineers use these techniques without ever thinking about the philosophical implications of this view.

Yet, as the concept has spread that probability distributions are the true "things" that science investigates, the philosophers and mathematicians have uncovered serious fundamental problems. I have examined some of these in passing in previous chapters. The next chapter is devoted to those problems.

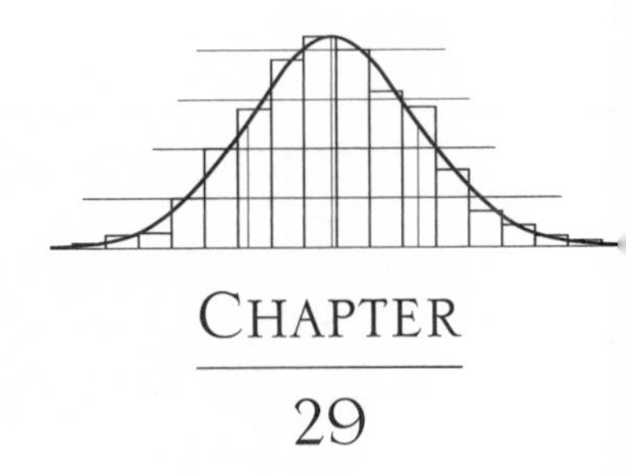

CHAPTER

29

THE IDOL WITH FEET OF CLAY

In 1962, Thomas Kuhn of the University of Chicago published *The Structure of Scientific Revolutions*. This book has had a profound influence on the way in which science is viewed by both philosophers and practitioners. Kuhn noted that reality is exceedingly complicated and can never be completely described by an organized scientific model. He proposed that science tends to produce a model of reality that appears to fit the data available and is useful for predicting the results of new experiments. Since no model can be completely true, the accumulation of data begins to require modifications of the model to correct it for new discoveries. The model becomes more and more complicated, with special exceptions and intuitively implausible extensions. Eventually, the model can no longer serve a useful purpose. At that point, original thinkers will emerge with an entirely different model, creating a revolution in science.

The statistical revolution was an example of this exchange of models. In the deterministic view of nineteenth-century science, Newtonian physics had effectively described the

An orrery, a clockwork mechanism designed to show the movement of the planets about the Sun

motion of planets, moons, asteroids, and comets—all of it based on a few well-defined laws of motion and gravity. Some success had been achieved in finding laws of chemistry, and Darwin's law of natural selection appeared to provide a useful start to understanding evolution. Attempts were even made to extend the search for scientific laws into the realms of sociology, political science, and psychology. It was believed at the time that the major problem with finding these laws lay in the imprecision of the measurements.

Mathematicians like Pierre Simon Laplace of the early nineteenth century developed the idea that astronomical measurements involve slight errors, possibly due to atmospheric conditions or to the human fallibility of the observers. He opened the door to the statistical revolution by proposing that these errors would have a probability distribution. To use Thomas Kuhn's view, this was a modification of the clockwork universe made necessary by new data. The nineteenth-century Belgian polymath Lambert Adolphe Jacques Quételet anticipated the statistical revolution by proposing that the laws of human behavior were probabilistic in nature. He did not have Karl Pearson's multiple parameter approach and was unaware of the need for optimum estimation methods, and his models were exceedingly naive.

Eventually, the deterministic approach to science collapsed because the differences between what the models predicted and what was actually observed grew greater with more precise measurements. Instead of eliminating the errors that Laplace thought were interfering with the ability to observe the true motion of the planets, more precise measurements showed more and more variation. At this point, science was ready for Karl Pearson and his distributions with parameters.

The preceding chapters of this book have shown how Pearson's statistical revolution has come to dominate all of modern science. In spite of the apparent determinism of molecular biology, where genes are found that cause cells to generate specific proteins, the actual data from this science are filled with randomness and the genes are, in fact, parameters of the distribution of those results. The effects of modern drugs on bodily functions, where doses of 1 or 2 milligrams cause profound changes in blood pressure or psychic neuroses, seem to be exact. But the pharmacological studies that prove these effects are designed and analyzed in terms of probability distributions, and the effects are parameters of those distributions.

Similarly, the statistical methods of econometrics are used to model the economic activity of a nation or of a firm. The subatomic particles we confidently think of as electrons and protons are described in quantum mechanics as probability distributions. Sociologists derive weighted sums of averages taken across populations to describe the interactions of individuals—but only in terms of probability distributions. In many of these sciences, the use of statistical models is so much a part of their methodology that the parameters of the distributions are spoken of as if they were real, measurable things. The uncertain conglomeration of shifting and changing measurements that are the starting point of these sciences is submerged in the calculations, and the conclusions are stated in terms of parameters that can never be directly observed.

THE STATISTICIANS LOSE CONTROL

So ingrained in modern science is the statistical revolution that the statisticians have lost control of the process. The probability calculations of the molecular geneticists have been developed independently of the mathematical statistical literature. The new discipline of information science has emerged from the ability of the computer to accumulate large amounts of data and the need to make sense out of those huge libraries of information. Articles in the new journals of information science seldom refer to the work of the mathematical statisticians, and many of the techniques of analysis that were examined years ago in *Biometrika* or the *Annals of Mathematical Statistics* are being rediscovered. The applications of statistical models to questions of public policy have spawned a new discipline called "risk analysis," and the new journals of risk analysis also tend to ignore the work of the mathematical statisticians.

Scientific journals in almost all disciplines now require that the tables of results contain some measure of the statistical uncertainty associated with the conclusions, and standard methods of statistical analysis are taught in universities as part of the graduate courses in these sciences, usually without involving the statistics departments that may exist at these same universities.

In the more than a hundred years since Karl Pearson's discovery of the skew distributions, the statistical revolution has not only been extended to most of science, but many of its ideas have spread into the general culture. When the television news anchor announces that a medical study has shown that passive smoking "doubles the risk of death" among nonsmokers, almost everyone who listens thinks he or she knows what that means. When a public opinion poll is used to declare that 65 percent of the public think the president is doing a good job, plus or minus 3 percent, most of us think we understand both the 65 percent and the 3 percent. When the weatherman predicts a 95 percent chance of rain tomorrow, most of us will take along an umbrella.

The statistical revolution has had an even more subtle influence on popular thought and culture than just on the way we bandy probabilities and proportions about as if we knew what they meant. We accept the conclusions of scientific investigations based upon estimates of parameters, even if none of the actual measurements agrees exactly with those conclusions. We are willing to make public policy and organize our personal plans using averages of masses of data. We take for granted that assembling data on deaths and births is not only a proper procedure but a necessary one, and we have no fear of angering the gods by counting people. On the level of language, we use the words *correlation* and *correlated* as if they mean something, and as if we think we know their meaning.

This book has been an attempt to explain to the nonmathematician something about this revolution. I have tried to describe the essential ideas behind the revolution, how it came to be adopted in a number of different areas of science, how it eventually came to dominate almost all of science. I have tried to interpret some of the mathematical models with words and examples that can be understood without having to climb the heights of abstract mathematical symbolism.

HAS THE STATISTICAL REVOLUTION RUN ITS COURSE?

The world "out there" is an exceedingly complicated mass of sensations, events, and turmoil. With Thomas Kuhn, I do not believe that the human mind is capable of organizing a structure of ideas that can come even close to describing what is really out there. Any attempt to do so contains fundamental faults. Eventually, those faults will become so obvious that the scientific model must be continuously modified and eventually discarded in favor of a more subtle one. We can expect the statistical revolution will eventually run its course and be replaced by something else.

It is only fair that I end this book with some discussion of the philosophical problems that have emerged as statistical methods have been extended into more and more areas of human endeavor. What follows will be an adventure in philosophy. The reader may wonder what philosophy has to do with science and real life. My answer is that philosophy is not some arcane academic exercise done by strange people called philosophers. Philosophy looks at the underlying assumptions behind our day-to-day cultural ideas and activities. Our worldview, which we learn from our culture, is governed by subtle assumptions. Few of us are even aware of them. The study of philosophy allows us to uncover these assumptions and examine their validity.

I once taught a course in the mathematics department of Connecticut College. The course had a formal name, but the department members referred to it as "math for poets." It was designed as a one-semester course to acquaint liberal arts majors with the essential ideas of mathematics. Early in the semester, I introduced the students to the *Ars Magna* of Girolamo Cardano, a sixteenth-century Italian mathematician. The *Ars Magna* contains the first published description of the emerging methods of algebra. Echoing his tome, Cardano writes in his introduction that this algebra is not new. He is no ignorant fool, he implies. He is aware that ever since the fall of man, knowledge has been decreasing and that Aristotle knew far more than anyone living at Cardano's time. He is aware that there can be no new knowledge. In his ignorance, however, he has been unable to find reference to a particular idea in Aristotle, and so he presents his readers with this idea, which appears to be new. He is sure that some more knowledgeable reader will locate where, among the writings of the ancients, this idea that appears to be new can actually be found.

The students in my class, raised in a cultural milieu that not only believes new things can be found but that actually encourages innovation, were shocked. What a stupid thing to write! I pointed out to them that in the sixteenth century, the worldview of Europeans was constrained by fundamental philosophical assumptions. An important part of their worldview was the idea of the fall of man and the subsequent continual deterioration of the world—of morals, of knowledge, of industry, of all things. This was known to be true, so true that it was seldom even articulated.

I asked the students what underlying assumptions of their worldview might possibly seem ridiculous to students 500 years in the future. They could think of none.

As the surface ideas of the statistical revolution spread through modern culture, as more and more people believe its truths without thinking about its underlying assumptions, let us consider three philosophical problems with the statistical view of the universe:

1. Can statistical models be used to make decisions?

2. What is the meaning of probability when applied to real life?

3. Do people really understand probability?

Can Statistical Models Be Used to Make Decisions?

L. Jonathan Cohen of Oxford University has been a trenchant critic of what he calls the "Pascalian" view, by which he means the use of statistical distributions to describe reality. In his 1989 book, *An Introduction to the Philosophy of Induction and Probability*, he proposes the paradox of the lottery, which he attributes to Seymour Kyberg of Wesleyan University in Middletown, Connecticut.

Suppose we accept the ideas of hypothesis or significance testing. We agree that we can decide to reject a hypothesis about reality if the probability associated with that hypothesis is very small. To be specific, let's set 0.0001 as a very small probability. Let's now organize a fair lottery with 10,000 numbered tickets. Consider the hypothesis that ticket number 1 will win the lottery. The probability of that is 0.0001. We reject that hypothesis. Consider the hypothesis that ticket number 2 will win the lottery. We can also reject that hypothesis. We can reject similar hypotheses for any specific numbered ticket. Under the rules of logic, if A is not true, and B is not true, and C is not true, then (A or B or C) is not true. That is, under the rules of logic, if every specific ticket will not win the lottery, then no ticket will win the lottery.

In an earlier book, *The Probable and the Provable*, L. J. Cohen proposed a variant of this paradox based on common legal practice. In the common law, a plaintiff in a civil suit wins if his claim seems true on the basis of the "preponderance" of the evidence. This has been accepted by the courts to mean that the probability of the plaintiff's claim is greater than 50 percent. Cohen proposes the paradox of the gate-crashers. Suppose there is a rock concert in a hall with 1,000 seats. The promoter sells tickets for 499 of the seats, but when the concert starts, all 1,000 seats are filled. Under English common law, the promoter has the right to collect from each of the 1,000 persons at the concert, since the probability that any one of them is a gate-crasher is 50.1 percent. Thus, the promoter will collect money from 1,499 patrons for a hall that holds only 1,000.

What both paradoxes show is that decisions based on probabilistic arguments are not logical decisions. Logic and probabilistic arguments are incompatible. R. A. Fisher justified inductive reasoning in science by appealing to significance tests based on well-designed experiments. Cohen's paradoxes suggest that such inductive reasoning is illogical. Jerry Cornfield justified the finding that smoking causes lung cancer by appealing to a piling up of evidence, where study after study shows results that are highly improbable unless you assume that smoking is the cause of the cancer. Is it illogical to believe that smoking causes cancer?

This lack of fit between logic and statistically based decisions is not something that can be accounted for by finding a faulty assumption in Cohen's paradoxes. It lies at the heart of what is meant by logic. (Cohen proposes that probabilistic models be replaced by a sophisticated version of mathematical logic known as "modal logic," but I think this solution introduces more problems than it solves.) In logic, there is a clear difference between a proposition that is true and one that is false. But probability introduces the idea that some propositions are probably or almost true. That little bit of resulting unsureness blocks our ability to apply the cold exactness of material implication in dealing with cause and effect. One of the solutions proposed for this problem in clinical research is to look upon each clinical study as providing some information about the effect of a given treatment. The value of that information can be determined by a statistical analysis of the study but also by the quality of the study. This extra measure, the quality of the study, is used to determine which studies will dominate in the conclusions. The concept of quality of a study is a vague one and not easily calculated. The paradox remains, eating at the heart of statistical methods. Will this worm of inconsistency require a new revolution in the twenty-first century?

What Is the Meaning of Probability When Applied to Real Life?

Andrei Kolmogorov established the mathematical meaning of probability: Probability is a measure of sets in an abstract space of events. All the mathematical properties of probability can be derived from this definition. When we wish to apply probability to real life, we need to identify that abstract space of events for the particular problem at hand. When the weather forecaster says that the probability of rain tomorrow is 95 percent, what is the set of abstract events being measured? Is it the set of all people who will go outside tomorrow, 95 percent of whom will get wet? Is it the set of all possible moments in time, 95 percent of which will find me getting wet? Is it the set of all one-inch-square pieces of land in a given region, 95 percent of which will get wet? Of course, it is none of those. What is it, then?

Karl Pearson, coming before Kolmogorov, believed that probability distributions were observable by just collecting a lot of data. We have seen the problems with that approach.

William S. Gosset attempted to describe the space of events for a designed experiment. He said it was the set of all possible outcomes of that experiment. This may be intellectually satisfying, but it is useless. It is necessary to describe the probability distribution of outcomes of the experiment in sufficient exactitude so we can calculate the probabilities needed to do statistical analysis. How does one derive a particular probability distribution from the vague idea of the set of all possible outcomes?

R. A. Fisher first agreed with Gosset, but then he developed a much better definition. In his experimental designs, treatments are assigned to units of experimentation at random. If we wish to compare two treatments for hardening of the arteries in obese rats, we randomly assign treatment A to some rats and treatment B to the rest. The study is run, and we observe the results. Suppose that both treatments have the same underlying effect. Since the animals were assigned to treatment at random, any other assignment would have produced similar results. The random labels of treatment are irrelevant tags that can be switched among animals—as long as the treatments have the same effect. Thus, for Fisher, the space of events is the set of all possible random assignments that could have been made. This is a finite set of events, all of them equally probable. It is possible to compute the probability distribution of the outcome under the null hypothesis that the treatments have the same effect. This is called the "permutation test." When Fisher proposed it, the counting of all possible random assignments was impossible. Fisher proved that his formulas for analysis of variance provided good approximations to the correct permutation test.

That was before the day of the high-speed computer. Now it is possible to run permutation tests (the computer is tireless when it comes to doing simple arithmetic), and Fisher's formulas for analysis of variance are not needed. Nor are many of the clever theorems of mathematical statistics that were proved over the years. All significance tests can be run with permutation tests on the computer, as long as the data result from a randomized controlled experiment.

When a significance test is applied to observational data, this is not possible. This is a major reason Fisher objected to the studies of smoking and health. The authors were using statistical significance tests to prove their case. To Fisher, statistical significance tests are inappropriate unless they are run in conjunction with randomized experiments. Discrimination cases in American courts are routinely decided on the basis of statistical significance tests. The U.S. Supreme Court has ruled that this is an acceptable way of determining if there was a disparate impact due to sexual or racial discrimination. Fisher would have objected loudly. In the late 1980s, the U.S. National Academy of Sciences sponsored a study of the use of statistical methods as evidence in courts. Chaired by Stephen Fienberg of Carnegie Mellon University and Samuel Krislov of the University of Minnesota, the study committee issued its report in 1988. Many of the papers included in that report criticized the use of hypothesis tests in discrimination cases, with arguments similar to those used by Fisher when he objected to the proof that smoking caused cancer. If the Supreme Court wants to approve significance tests in litigation, it should identify the space of events that generate the probabilities.

A second solution to the problem of finding Kolmogorov's space of events occurs in sample survey theory. When we wish to take a random sample of a population to determine something about it, we identify exactly the population of people to be examined, establish a method of selection, and sample at random according to that method. There is uncertainty in the conclusions, and we can apply statistical methods to quantify that uncertainty.

The uncertainty is due to the fact that we are dealing with a sample of the populace. The true values of the universe being examined, such as the true percentage of American voters who approve of the president's policies, are fixed. They are just not known. The space of events that enables us to use statistical methods is the set of all possible random samples that might have been chosen. Again, this is a finite set, and its probability distribution can be calculated. The real-life meaning of probability is well established for sample surveys.

It is not well established when statistical methods are used for observational studies in astronomy, sociology, epidemiology, law, or weather forecasting. The disputes that arise in these areas are often based on the fact that different mathematical models will give rise to different conclusions. If we cannot identify the space of events that generate the probabilities being calculated, then one model is no more valid than another. As has been shown in many court cases, two expert statisticians working with the same data can disagree about the analyses of those data. As statistical models are used more and more for observational studies to assist in social decisions by government and advocacy groups, this fundamental failure to be able to derive probabilities without ambiguity will cast doubt on the usefulness of these methods.

Do People Really Understand Probability?

One solution to the question of the real-life meaning of probability has been the concept of "personal probability." L. J. ("Jimmie") Savage of the United States and Bruno de Finetti of Italy were the foremost proponents of this view. This position was best presented in Savage's 1954 book, *The Foundations of Statistics*. In this view, probability is a widely held concept. People naturally govern their lives using probability. Before entering on a venture, people intuitively decide on the probability of possible outcomes. If the probability of danger, for instance, is thought to be too great, a person will avoid that action. To Savage and de Finetti, probability is a common concept. It does not need to be connected with Kolmogorov's mathematical probability. All we need do is establish general rules for making personal probability coherent. To do this, we need only to assume that people will not be inconsistent when judging the probability of events. Savage derived rules for internal coherence based upon that assumption.

Under the Savage–de Finetti approach, personal probability is unique to each person. It is perfectly possible that one person will decide that the probability of rain is 95 percent and that another will decide that it is 72 percent—based on their both observing the same data. Using Bayes's theorem, Savage and de Finetti were able to show that two people with coherent personal probabilities will converge to the same final estimates of probability if faced with a sequence of the same data. This is a satisfying conclusion. People are different but reasonable, they seemed to say. Given enough data, such reasonable people will agree in the end, even if they disagreed to begin with.

John Maynard Keynes in his doctoral thesis, published in 1921 as *A Treatise on Probability*, thought of personal probability as something else. To Keynes, probability was the measure of uncertainty that all people with a given cultural education would apply to a given situation. Probability was the result of one's culture, not just of one's inner, gut feeling. This approach is difficult to support if we are trying to decide between a probability of 72 percent and one of 68 percent. General cultural agreement could never reach such a degree of precision. Keynes pointed out that for decision making we seldom, if ever, need to know the exact numerical probability of some event. It is usually sufficient to be able to order events. According to Keynes, decisions can be made from knowing that it is more probable that it will rain tomorrow than that it will hail, or that it is twice as probable that it will rain as that it will hail. Keynes points out that probability can be a partial ordering. One does not have to compare everything with everything else. One can ignore the probability relationships between whether the Yankees will win the pennant and whether it will rain tomorrow.

In this way, two of the solutions to the problem of the meaning of probability rest on the general human desire to quantify uncertainty, or at least to do so in a rough fashion. In his *Treatise*, Keynes works out a formal mathematical structure for his partial ordering of personal probability. He did this work before Kolmogorov laid the foundations for mathematical probability, and there is no attempt to link his formulas to Kolmogorov's work. Keynes claimed that his definition of probability was different from the set of mathematical counting formulas that represented the mathematics of probability in 1921. For Keynes's probabilities to be usable, the person who invokes them would still have to meet Savage's criteria for coherence.

This makes for a view of probability that might provide the foundations for decision making with statistical models. This is the view that probability is not based on a space of events but that probabilities as numbers are generated out of the personal feelings of the people involved. Then psychologists Daniel Kahneman and Amos Tversky of Hebrew University in Jerusalem began their investigations of the psychology of personal probability.

Through the 1970s and 1980s, Kahneman and Tversky investigated the way in which individuals interpret probability. Their work was summed up in their book (coedited by P. Slovic) *Judgment Under Uncertainty: Heuristics and Biases*. They presented a series of probabilistic scenarios to college students, college faculty, and ordinary citizens. They found no one who met Savage's criteria for coherence. They found, instead, that most people did not have the ability to keep even a consistent view of what different numerical probabilities meant. The best they could find was that people could keep a consistent sense of the meaning of 50:50 and the meaning of "almost certain." From Kahneman and Tversky's work, we have to conclude that the weather forecaster who tries to distinguish between a 90 percent probability of rain and a 75 percent probability of rain cannot really tell the difference. Nor do any of the listeners to their forecast have a consistent view of what that difference means.

In 1974, Tversky presented these results at a meeting of the Royal Statistical Society. In the discussion afterward, Patrick Suppes of Stanford University proposed a simple probability model that met Kolmogorov's axioms and that also mimicked what Kahneman and Tversky had found. This means that people who used this model would be coherent in their personal probabilities. In Suppes's model, there are only five probabilities:

surely true

more probable than not

as probable as not

less probable than not

surely false

This leads to an uninteresting mathematical theory. Only about a half-dozen theorems can be derived from this model, and their proofs are almost self-evident. If Kahneman and Tversky are right, the only useful version of personal probability provides no grist at all for the wonderful abstractions of mathematics, and it generates the most limited versions of statistical models. If Suppes's model is, in fact, the only one that fits personal probability, many of the techniques of statistical analysis that are standard practice are useless, since they only serve to produce distinctions below the level of human perception.

Is Probability Really Necessary?

The basic idea behind the statistical revolution is that the real things of science are distributions of numbers, which can be described by parameters. It is mathematically convenient to embed that concept into probability theory and deal with probability distributions. By considering the distributions of numbers as elements from the mathematical theory of probability, it is possible to establish optimum criteria for estimators of those parameters and to deal with the mathematical problems that arise when data are used to describe the distributions. Because probability seems inherent in the concept of a distribution, much effort has been spent on getting people to understand probability, trying to link the mathematical idea of probability to real life, and using the tools of conditional probability to interpret the results of scientific experiments and observations.

The idea of distribution can exist outside of probability theory. In fact, improper distributions (which are improper because they do not meet all the requirements of a probability distribution) are already being used in quantum mechanics and in some Bayesian techniques. The development of queuing theory, a situation where the average time between arrivals at the queue equals the average service time in the queue, leads to an improper distribution for the amount of time someone entering the queue will have to wait. Here is a case where applying the mathematics of probability theory to a real-life situation leads us out of the set of probability distributions.

WHAT WILL HAPPEN IN THE TWENTY-FIRST CENTURY?

Kolmogorov's final insight was to describe probability in terms of the properties of finite sequences of symbols, where information theory is not the outcome of probabilistic calculations but the progenitor of probability itself. Perhaps someone will pick up the torch where he left it and develop a new theory of distributions where the very nature of the digital computer is brought into the philosophical foundations.

Who knows where there may be another R. A. Fisher, working on the fringes of established science, who will soon burst upon the scene with insights and ideas that have never been thought before? Perhaps, somewhere in the center of China, another Lucien Le Cam has been born to an illiterate farm family; or in North Africa, another George Box who stopped his formal education after secondary school may now be working as a mechanic, exploring and learning on his own. Perhaps another Gertrude Cox will soon give up her hopes of being a missionary and become intrigued with the puzzles of science and mathematics; or another William S. Gossett is trying to find a way to solve a problem in the brewing of beer; or another Neyman or Pitman is teaching in some obscure provincial college in India and thinking deep thoughts. Who knows from where the next great discovery will come?

As we enter the twenty-first century, the statistical revolution in science stands triumphant. It has vanquished determinism from all but a few obscure corners of science. It has become so widely used that its underlying assumptions have become part of the unspoken popular culture of the Western world. It stands triumphant on feet of clay. Somewhere, in the hidden corners of the future, another scientific revolution is waiting to overthrow it, and the men and women who will create that revolution may already be living among us.

AFTERWORD

By writing this book, I have divided the women and men who have contributed to the field into two groups: those whom I mention and those whom I do not mention. The first group may object to the fact that I have tended to describe only a small part of their work. The second group may object that I have mentioned nothing at all about their work. A decent respect for the feelings of both groups requires that I explain my methods of choosing what to describe and what to leave out.

The first group of omissions is due to the fact that modern science has become too extensive for any one person to know about all its ramifications. As a result, there are areas of research in which statistical methods have been used but of which I am, at best, barely aware. In the early 1970s, I ran a literature search on the use of computers in medical diagnoses. I found three independent traditions. Within any one of them, the workers all referred to one another and published in the same journals. There was no indication that any of the scientists in each of these groups had the slightest knowledge of the work being done by the other groups. This was within the relatively small world of medicine. In the wider range of science in general, there are probably groups using statistical methods and publishing in journals that I have never heard of. My knowledge of the statistical revolution results from my reading in the mainstream of mathematical statistics. Scientists who do not read or contribute to the journals I read, like the engineers developing fuzzy set theory, may be doing notable work, but if they do not publish in the scientific/mathematical tradition of which I am aware, I could not include them.

There are omissions even of material I do know about. I did not set out to write a comprehensive history of the development of statistical methodology. Since this book is aimed at readers with little or no mathematical training, I had to choose examples that I could explain in words, without the use of mathematical symbols. This further limited the choice of individuals whose work I could describe. I also wanted to keep a sense of connection running through the book. If I had had the use of mathematical notation, I could have shown the relationships among a large collection of topics. But without that notation, the book could easily have degenerated into a collection of ideas that do not seem to have any connection. This book needed a path of somewhat organized topics. The path I have chosen through the immense complications of twentieth-century statistics may not be the path that others would choose. Once chosen, it forced me to ignore many other aspects of statistics, as interesting as they might be to me.

The fact that I have left someone out of my book does not mean that that person's work is unimportant, or even that I think it unimportant. It only means that I could not find a way to include that work in this book as its structure developed.

I hope some readers will be inspired by this book to look into the statistical revolution more thoroughly. It is my hope that a reader may even be inspired to study the subject and join the world of statistical research. In the bibliography, I have isolated a small group of books and articles that I think are accessible to the nonmathematician. In these works, other statisticians have tried to explain what excites them about statistics. Those readers who would like to examine the statistical revolution further will want to read some of these.

I wish to acknowledge the efforts of those at W. H. Freeman who were instrumental in producing the final polished version of this book. I am indebted to Don Gecewicz for a thorough job of fact checking and editing; to Eleanor Wedge and Vivien Weiss for the final copyediting (and further fact checking); to Patrick Farace who saw the potential value of this book; and to Victoria Tomaselli, Bill Page, Karen Barr, Meg Kuhta, and Julia DeRosa for the artistic and production components of the effort.

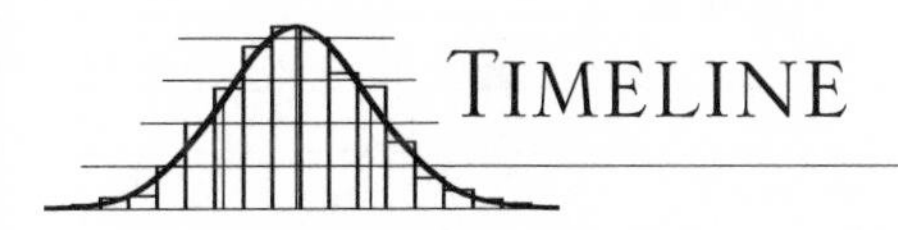

Timeline

Year	Event	Person
1857	Birth of Karl Pearson	K. Pearson
1865	Birth of Guido Castelnuovo	G. Castelnuovo
1866	Gregor Mendel's work on plant inbreeding	G. Mendel
1875	Birth of Francesco Paolo Cantelli	F. P. Cantelli
1876	Birth of William Sealy Gosset	W. S. Gosset ("Student")
1886	Birth of Paul Lévy	P. Lévy
1890	Birth of Ronald Aylmer Fisher	R. A. Fisher
1893	Birth of Prasanta Chandra Mahalanobis	P. C. Mahalanobis
1893	Birth of Harald Cramér	H. Cramér
1894	Birth of Jerzy Neyman	J. Neyman
1895	Discovery of skew distributions	K. Pearson
1895	Birth of Egon S. Pearson	E. S. Pearson
1899	Birth of Chester Bliss	C. Bliss
1900	Birth of Gertrude M. Cox	G. M. Cox
1900	Rediscovery of Gregor Mendel's work	W. Bateson
1902	First issue of *Biometrika*	F. Galton, K. Pearson, R. Weldon
1903	Birth of Andrei Nikolaevich Kolmogorov	A. N. Kolmogorov
1906	Birth of Samuel S. Wilks	S. S. Wilks
1908	"The Probable Error of the Mean" ("Student"'s t-test)	W. S. Gossett
1909	Birth of Florence Nightingale David	F. N. David
1911	Death of Sir Francis Galton	F. Galton
1911	*The Grammar of Science*	K. Pearson
1912	Birth of Jerome Cornfield	J. Cornfield
1912	R. A. Fisher's first publication	R. A. Fisher
1915	The distribution of the correlation coefficient	R. A. Fisher
1915	Birth of John Tukey	J. Tukey
1916	Glivenko-Cantelli lemma first appears	F. P. Cantelli
1917	Birth of L. J. ("Jimmie") Savage	L. J. Savage
1919	Publication of *Calcolo della probabilità* . . .	G. Castelnuovo
1919	Fisher at Rothamsted Experimental Station	R. A. Fisher
1920	First of the papers on Lebesgue integration	H. Lebesgue
1921	Treatise on Probability	J. M. Keynes
1921	"Studies in Crop Variation. I"	R. A. Fisher
1923	"Studies in Crop Variation. II"	R. A. Fisher

Year	Event	Person
1924	"Studies in Crop Variation. III"	R. A. Fisher
1924	"The Elimination of Mental Defect"—Fisher's first article on eugenics	R. A. Fisher
1925	First edition of *Statistical Methods for Research Workers*	R. A. Fisher
1925	The Theory of Statistical Estimation (ML Estimation)	R. A. Fisher
1926	First paper on experimental design in agriculture	R. A. Fisher
1927	"Studies in Crop Variation. IV"	R. A. Fisher
1928	The first of the Neyman-Pearson papers on hypothesis testing	J. Neyman, E. S. Pearson
1928	The three asymptotes of the extreme	L. H. C. Tippett, R. A. Fisher
1928	"Studies in Crop Variation. VI"	R. A. Fisher
1930	First issue of the *Annals of Mathematical Statistics*	H. Carver
1930	*The Genetical Theory of Natural Selection*	R. A. Fisher
1931	Founding of Indian Statistical Institute	P. C. Mahalanobis
1933	Axiomization of probability	A. N. Kolmogorov
1933	First issue of *Sankhya*	P. C. Mahalanobis
1933	Completion of work on probit analysis	C. Bliss
1933	Samuel S. Wilks arrives at Princeton	S. S. Wilks
1934	Neyman's confidence intervals	J. Neyman
1934	Proof of the central limit theorem	P. Lévy, J. Lindeberg
1934	Chester Bliss at Leningrad Institute for Plant Protection	C. Bliss
1935	First development of martingale theory	P. Lévy
1935	Publication of *The Design of Experiments*	R. A. Fisher
1936	Death of Karl Pearson	K. Pearson
1937	Enumerative check census for U.S. unemployment using random sampling	M. Hansen, F. Stephan
1937	Death of William Sealy Gosset	W. S. Gosset ("Student")
1938	*Statistical Tables for Biological, Agricultural, and Medical Research*	R. A. Fisher, F. Yates
1940	*Statistical Methods* textbook	G. W. Snedecor
1941	Death of Henri Lebesgue	H. Lebesgue
1945	Redaction of Fisher's work in *Mathematical Methods of Statistics*	H. Cramér
1945	Wilcoxon's first publication on nonparametric tests	F. Wilcoxon
1947	First appearance of sequential estimation theory in public domain	A. Wald

Year	Event	Person
1947	Mann-Whitney formulation of nonparametric tests	H. G. Mann, D. R. Whitney
1948	Pitman's work on nonparametric statistical inference	E. J. G. Pitman
1949	Cochran's work on observational studies	W. G. Cochran
1950	Publication of Cochran and Cox's book on experimental design	W. G. Cochran, G. M. Cox
1952	Death of Guido Castelnuovo	G. Castelnuovo
1957	Fisher's polemics about the supposed dangers of cigarette smoking	R. A. Fisher
1958	Publication of *Statistics of Extremes*	E. J. Gumbel
1959	Box applies the word "robust"	G. E. P. Box
1959	Definitive formulation of hypothesis testing	E. L. Lehmann
1960	*Combinatorial Chance*	F. N. David, D. E. Burton
1962	Formulation of Savage-de Finetti theory of personal probability	L. J. Savage, B. de Finetti
1962	Fisher's final paper deals with sex differences in genetics	R. A. Fisher
1962	Death of Ronald Aylmer Fisher	R. A. Fisher
1964	Death of Samuel S. Wilks	S. S. Wilks
1964	"An analysis of transformations"	G. E. P. Box, D. R. Cox
1966	Death of Francesco Paolo Cantelli	F. P. Cantelli
1967	Hájek's formulation of rank tests	J. Hájek
1969	National halothane study (including work on log-linear models)	Y. M. M. Bishop and others
1970	Nancy Mann's first publication on reliability theory and Weibull distribution	N. Mann
1970	*Games, Gods, and Gambling*	F. N. David
1971	Death of Paul Lévy	P. Lévy
1971	Death of L. J. ("Jimmie") Savage	L. J. Savage
1972	Princeton Study of Robust Estimation (Princeton Robustness Study)	D. F. Andrews, P. J. Bickel, F. R. Hampel, P. J. Huber, W. H. Rogers, J. W. Tukey
1972	Death of Prasanta Chandra Mahalanobis	P. C. Mahalanobis
1975	Stella Cunliffe elected president of the Royal Statistical Society	S. V. Cunliffe
1976	"Science and Statistics," a view of the uses of significance testing	G. E. P. Box
1977	Cox's formulation of significance testing	D. R. Cox
1977	Publication of *Exploratory Data Analysis*	J. Tukey
1978	Death of Gertrude M. Cox	G. M. Cox

Year	Event	Person
1979	Death of Chester Bliss	C. Bliss
1979	Death of Jerome Cornfield	J. Cornfield
1979	Janet Norwood named Commissioner of the Bureau of Labor Statistics	J. Norwood
1980	Death of Egon S. Pearson	E. S. Pearson
1981	Death of Jerzy Neyman	J. Neyman
1982	Modern formulation of chaos theory	R. Abraham, C. Shaw
1983	Studies showing the limited nature of personal probability	A. Tversky, D. Kahneman
1985	Death of Harald Cramér	H. Cramér
1987	Death of Andrei Nikolaevich Kolmogorov	A. N. Kolmogorov
1987	Application of kernel-based regression to focus cameras ("fuzzy systems")	T. Yamakawa
1989	L. J. Cohen's critique of statistical models and methods	L. J. Cohen
1990	*Spline Models for Observational Data*	G. Wahba
1992	Full development of martingale approach to medical studies	O. Aalen, E. Anderson, R. Gill
1995	Death of Florence Nightingale David	F. N. David
1997	Extension of Cochran's methods to sequential analysis	C. Jennison, B. W. Turnbull
1999	EM algorithm adapted to a problem involving the Aalen-Anderson-Gill martingale model	R. A. Betensky, J. C. Lindsey, L. M. Ryan
2000	Death of John Tukey	J. Tukey

Bibliography

Books and articles that are accessible to readers without mathematical training:

Boen, James R., and Zahn, Douglas A. 1994. *The Human Side of Statistical Consulting*. Belmont, CA: Lifetime Learning Publications. Boen and Zahn sum up their combined experiences as statistical consultants to scientists at a university (Boen) and in industry (Zahn). The book was written for statisticians entering the profession, but it is primarily a book about the psychological relationships among collaborating scientists. The insights and the examples used provide the reader with a very real feel for what the work of a consulting statistician is all about.

Box, George E. P. 1976. Science and statistics. *Journal of the American Statistical Association* 71:791. This is an address by George Box, in which he lays out his own philosophy of experimentation and scientific inference. Most of the material is accessible to readers without mathematical training.

Box, Joan Fisher. 1978. *R. A. Fisher, the Life of a Scientist*. New York: John Wiley & Sons. Joan Fisher Box is R. A. Fisher's daughter. In this biography of her father, she does an excellent job of explaining the nature and importance of much of his research. She also provides a view of him as a man, including personal reminiscences. She does not gloss over the less admirable aspects of his behavior (such as the time he abandoned his family) but shows an understanding of his motives and ideas.

Deming, W. Edwards. 1982. *Out of the Crisis*. Cambridge, MA: Massachusetts Institute of Technology, Center for Advanced Engineering Study. This is Deming's carefully written attempt to influence management of American companies. Without using mathematical notation, he explains important ideas like operational definition and sources of variance. He gives examples of situations in different industries. Above all, the book is an extension of this remarkable man; it reads exactly as he talked. He does not hold back on his criticism of management or of many of what he considered to be foolish aspects of American management practices, both in industry and government.

EFRON, BRADLEY. 1984. The art of learning from experience. *Science* 225:156. This short article explains the development of "the bootstrap" and other forms of computer-intensive resampling, written by the man who invented the bootstrap.

FISHER, R. A. 1956. *Statistical Methods and Scientific Inference.* Edinburgh: Oliver and Boyd. While it contains some mathematical derivations in its later chapters, this book was Fisher's attempt to explain what he meant by scientific inference, in carefully written words. It is his answer to work by Jerzy Neyman. Much of the material in this book appeared in earlier articles, but this is a summing up, by the genius who laid the foundations of modern mathematical statistics, of his views on what it all means.

HOOKE, R. 1983. *How to Tell the Liars From the Statisticians.* New York: Marcel Dekker. From time to time, statisticians, disturbed by the misuse of statistical methods in popular journals, have attempted to explain the concepts and procedures of good statistical practice to nonstatisticians. Unfortunately, it has been my observation that these books are read primarily by statisticians and are ignored by the people at whom they are aimed. This is one of the best of these books.

KOTZ, SAMUEL. 1965. Statistical terminology—Russian vs. English—in the light of the development of statistics in the U.S.S.R. *American Statistician* 19:22. Kotz was one of the first English-speaking statisticians to examine the work of the Russian school. He learned Russian to become a major translator of that work. In this article, he describes the peculiarities of Russian words as they are used in mathematical articles. The article also contains a detailed description of the fate of statistical methodology in the face of communist orthodoxy.

MANN, NANCY R. 1987. *The Keys to Excellence—The Story of the Deming Philosophy.* Los Angeles, CA: Preswick Books. Nancy Mann was head of the mathematical services groups at several West Coast industrial firms, became a member of the faculty at University of California, Los Angeles, and now heads a small consulting firm. Her contributions to the development of mathematical statistics include some extremely clever methods for estimating the parameters of a complicated class of distributions that are used in life testing of equipment. She had a great deal of contact with W. Edwards Deming and was one of his good friends. This is her explanation of Deming's work and methods for the nonmathematician.

PEARSON, KARL. 1911. *The Grammar of Science*. Meridian Library, NY: Meridian Library Edition (1957). Although some of the examples Pearson used are now outdated, having been superseded by new scientific discoveries, the insights and the bits of well-written philosophy in this book make it a delight to read almost 100 years later. It provides the reader with an excellent example of Pearson's style of writing and thought.

RAO, C. R. 1989. *Statistics and Truth: Putting Chance to Work*. Fairland, MD: International Co-Operative Publishing House. C. Radhakrishna Rao is one of the more honored members of the statistical profession. In his native land, he has been named Nehru Distinguished Professor and has been granted honorary doctorates from several Indian universities. A recipient of the American Statistical Association's Wilks Medal, he has been named a fellow of each of the four major statistical societies. Much of his published work involves extremely complicated derivations in multidimensions, but this book is the result of a series of popular lectures he gave in India. It presents his carefully thought-out concepts of the value, purpose, and philosophical ideas behind statistical modeling.

TANUR, JUDITH M., ed. 1972. *Statistics: A Guide to the Unknown*. San Francisco: Holden-Day, Inc. For the past twenty years, the American Statistical Association has had a program reaching out to high school students and college undergraduates. Committees of the association have prepared teaching materials, and the association has sponsored video tapes, several of which have appeared on public television. This book consists of a group of case studies, where statistical methods were applied to important social or medical problems. The case studies are written by statisticians who participated in them, but they are written to be read and understood by a high school student with little or no mathematics background. There are forty-five essays in this book, each running about ten pages long. The authors include some of the people I have mentioned in this book. Some of the questions attacked are: Can people postpone their deaths? and Does an increase in police manpower decrease the incidence of crime? Other topics include a discussion of close elections, a brief description of the analysis of the *Federalist* papers, how new food products are evaluated, the consumer price index, predicting future population growth, cloud seeding experiments, and the aiming of antiaircraft fire.

TUKEY, JOHN W. 1977. *Exploratory Data Analysis*. Reading, MA: Addison-Wesley Publishing Company. This is the textbook Tukey wrote for first-year statistics students at Princeton University. It assumes no prior knowledge—not even of high school algebra. It approaches statistical reasoning from the standpoint of a person faced with a set of data.

Collected works of prominent statisticians:

BOX, GEORGE E. P. 1985. *The Collected Works of George E. P. Box*. Belmont, CA: Wadsworth Publishing Company.

COCHRAN, W. G. 1982. *Contributions to Statistics*. New York: John Wiley & Sons.

FIENBERG, S. E., Hoaglin, D. C., Kruskal, W. H., and Tanur, J. M., eds. 1990. *A Statistical Model: Frederick Mosteller's Contributions to Statistics, Science, and Public Policy*. New York: Springer-Verlag.

FISHER, R. A. 1971. *Collected Papers of R. A. Fisher*. Edited by J. H. Bennett. Adelaide: The University of Adelaide.

——. R. A. 1950. *Contributions to Mathematical Statistics*. New York: John Wiley & Sons.

GOSSET, WILLIAM SEALY. 1942. *"Student" 's Collected Papers*. Edited by E. S. Pearson and John Wishart. Cambridge: Cambridge University Press.

NEYMAN, JERZY. 1967. *A Selection of Early Statistical Papers of J. Neyman*. Berkeley, CA: University of California Press.

NEYMAN, J., and Kiefer, J. 1985. *Proceedings of the Berkeley Conference in Honor of Jerzy Neyman and Jack Kiefer*. Edited by Lucien M. Le Cam and R. A. Olshen. Monterey, CA: Wadsworth Advanced Books.

SAVAGE, L. J. 1981. *The Writings of Leonard Jimmie Savage—A Memorial Selection*. Washington, DC: The American Statistical Association and the Institute of Mathematical Statistics.

TUKEY, J. W. 1984. *The Collected Works of John W. Tukey*. Edited by W. S. Cleveland. Belmont, CA: Wadsworth Advanced Books.

Obituaries, reminiscences, and published conversations:

ALEXANDER, KENNETH S. 1996. A conversation with Ted Harris. *Statistical Science* 11:150.

ANDERSON, R. L. 1980. William Gemmell Cochran, 1909–1980: A personal Tribute. *Biometrics* 36:574.

ANDERSON, T. W. 1996. R. A. Fisher and multivariate analysis. *Statistical Science* 11:20.

ANDREI NIKOLAEVICH KOLMOGOROV: 1903–1987. *IMS Bulletin* 16:324.

ANSCOMBE, FRANCIS J., moderator. 1988. Frederick Mosteller and John W. Tukey: A conversation. *Statistical Science* 3:136.

ARMITAGE, PETER. 1997. The Biometric Society—50 years on. *Biometric Society Newsletter*, 3.

——. 1977. A tribute to Austin Bradford Hill. *Journal of the Royal Statistical Society, Series* A 140:127.

BANKS, DAVID L. 1996. A conversation with I. J. Good. *Statistical Science* 11:1.

BARNARD, G. A., and Godambe, V. P. 1982. Memorial article, Allan Birnbaum, 1923–1976. *The Annals of Statistics* 10:1033.

BLOM, GUNNAR. 1987. Harald Cramér, 1893–1985. *Annals of Statistics* 15:1335.

BOARDMAN, THOMAS J. 1994. The statistician who changed the world: W. Edwards Deming, 1900–1993. *American Statistician* 48:179.

CAMERON, J. M., and Rosenblatt, J. R. 1995. Churchill Eisenhart, 1913–1994. *IMS Bulletin* 24:4.

CHESTER ITTNER BLISS, 1899–1979. 1979. *Biometrics* 35:715.

CRAIG, CECIL C. 1978. Harry C. Carver, 1890–1977. *Annals of Statistics* 6:1.

CUNLIFFE, STELLA, V. 1976. Interaction, the address of the president, delivered to the Royal Statistical Society on Wednesday, November 12, 1975. *Journal of the Royal Statistical Society, Series* A 139:1.

DANIEL, C., and Lehmann, E. L. 1979. Henry Scheffé, 1907–1977. *Annals of Statistics* 7:1149.

DARNELL, ADRIAN C. 1988. Harold Hotelling, 1895–1973. *Statistical Science* 3:57.

DAVID, HERBERT A. 1981. Egon S. Pearson, 1895–1980. *American Statistician* 35:94.

DEGROOT, MORRIS H. 1987. A conversation with George Box. *Statistical Science* 2:239.

——. 1986. A conversation with David Blackwell. *Statistical Science* 1:40.

——. 1986. A conversation with Erich L. Lehmann. *Statistical Science* 1:243.

——. 1986. A conversation with Persi Diaconis. *Statistical Science* 1:319.

DEMING, W. EDWARDS. 1972. P. C. Mahalanobis (1893–1972). *American Statistician* 26:49.

DPAC, VACLAV. 1975. Jaraslav Hájek, 1926–1974. *Annals of Statistics* 3:1031.

FIENBERG, STEPHEN E. 1994. A conversation with Janet L. Norwood. *Statistical Science* 9:574.

FRANKEL, MARTIN, and King, Benjamin. 1996. A conversation with Leslie Kish. *Statistical Science* 11:65.

GALTON, FRANCIS, F. R. S. 1988. Men of science, their nature and their nurture: Report of a lecture given Friday evening, 27 February 1874, at the Royal Institutions, taken from *Nature*, 5 March, 1874, pp. 344–345. *IMS Bulletin* 17:280.

GANI, J., ed. 1982. *The Making of Statisticians*. New York: Springer-Verlag.

GEISSER, SEYMOUR. 1986. Opera Selecta Boxi. *Statistical Science* 1:106.

GLADYS I. PALMER, 1895–1967. 1967. *American Statistician* 21:35.

GREENHOUSE, SAMUEL W., and Halperin, Max. 1980. Jerome Cornfield, 1912–1979. *American Statistician* 34:106.

GRENANDER, ULF, ed. 1959. *Probability and Statistics: The Harald Cramér Volume*. Stockholm: Almqvist and Wiksell.

HANSEN, MORRIS H. 1987. Some history and reminiscences on survey sampling. *Statistical Science* 2:180.

HEYDE, CHRIS. 1995. A conversation with Joe Gani. *Statistical Science* 10:214.

JEROME CORNFIELD'S publications. 1982. *Biometrics Supplement* 47.

JERRY CORNFIELD, 1912–1979. 1980. *Biometrics* 36:357.

KENDALL, DAVID G. 1991. Kolmogorov as I remember him. *Statistical Science* 6:303.

——. 1970. Ronald Aylmer Fisher, 1890–1902. In *Studies in the History of Statistics and Probability*, edited by E. S. Pearson and M. Kendall, 439. London: Hafner Publishing Company.

KUEBLER, ROY R. 1988. Raj Chandra Bose: 1901–1987. *IMS Bulletin* 17:50.

LAIRD, NAN M. 1989. A conversation with F. N. David. *Statistical Science* 4:235.

LE CAM, L. 1986. The central limit theorem around 1935. *Statistical Science* 1:78.

LEDBETTER, ROSS. 1995. Stamatis Cambanis, 1943–1995. *IMS Bulletin* 24:231.

LEHMANN, ERIC. L. 1997. Testing statistical hypotheses: The story of a book. *Statistical Science* 12:48.

LINDLEY, D. V. 1980. L. J. Savage—His Work in Probability and Statistics. *Annals of Statistics* 8:1.

LOEVE, MICHEL. 1973. Paul Lévy, 1886–1971. *Annals of Probability* 1:1.

MAHALANOBIS, P. C. 1938. Professor Ronald Aylmer Fisher, early days. *Sankhya* 4:265.

MONROE, ROBERT J. 1980. Gertrude Mary Cox, 1900–1978. *American Statistician* 34:48.

MUKHOPADHYAY, MITIS. 1997. A conversation with Sujit Kumar Mitra. *Statistical Science* 12:61.

NELDER, JOHN. 1994. Frank Yates: 1902–1994. *IMS Bulletin* 23:529.

NEYMAN, JERZY. 1981. Egon S. Pearson (August 11, 1895–June 12, 1980), an appreciation. *Annals of Statistics* 9:1.

OLKIN, INGRAM. 1989. A conversation with Maurice Bartlett. *Statistical Science* 4:151.

———. 1987. A conversation with Morris Hansen. *Statistical Science* 2:162.

ORD, KEITH. 1984. In memoriam, Maurice George Kendall, 1907–1983. *American Statistician* 38:36.

PEARSON, E. S. n.d. *The Neyman-Pearson Story: 1926–34*. Research Papers in Statistics. London: University College.

———. 1968. Studies in the history of probability and statistics. XX: Some early correspondence between W. S. Gosset, R. A. Fisher, and Karl Pearson, with notes and comments. *Biometrika* 55:445.

RADE, LENNART. 1997. A conversation with Harald Bergstrom. *Statistical Science* 12:53.

RAO, C. RADHAKRISHNA. 1993. Prasanta Chandra Mahalanobis, June 29, 1893–June 28, 1972. *IMS Bulletin* 22:593.

THE REVEREND THOMAS BAYES, F. R. S., 1701–1761. 1988. *IMS Bulletin* 17:276.

SAMUEL-CAHN, ESTER. 1992. A conversation with Esther Seiden. *Statistical Science* 7:339.

SHIRYAEV, A. N. 1991. Everything about Kolmogorov was unusual. *Statistical Science* 6:313.

SMITH, ADRIAN. 1995. A conversation with Dennis Lindley. *Statistical Science* 10:305.

SMITH, WALTER L. 1978. Harold Hotelling, 1895–1973. *Annals of Statistics* 6:1173.

STEPHAN, R. R., Tukey, J. W., Mosteller, F., Mood, A. M., Hansen, M. H., Simon, L. E., and Dixon, W. J. 1965. Samuel S. Wilks. *Journal of the American Statistical Association* 60:939.

STIGLER, STEPHEN M. 1989. Francis Galton's account of the invention of correlation. *Statistical Science* 4:73.

——. 1977. Eight centuries of sampling inspection: The trial of the pyx. *Journal of the American Statistical Association* 72:493.

STINNETT, SANDRA, et al. 1990. Women in statistics: Sesquicentennial activities. *American Statistician* 44:74.

STRAF, MIRON, and Olkin, Ingram. 1994. A conversation with Margaret Martin. *Statistical Science* 9:127.

SWITZER, PAUL. 1992. A conversation with Herbert Solomon. *Statistical Science* 7:388.

TAYLOR, G. I. 1973. Memories of Von Karman. *SIAM Review* 15:447.

TAYLOR, WALLIS. 1977. Lancelot Hogben, F.R.S. (1895–1975). *Journal of the Royal Statistical Society, Series* A, Part 2:261.

TEICHROEW, DANIEL. 1965. A history of distribution sampling prior to the era of the computer and its relevance to simulation. *Journal of the American Statistical Association* 60:27.

WATSON, G. S. 1982. William Gemmell Cochran, 1909–1980. *Annals of Statistics* 10:1.

WHITNEY, RANSOM. 1997. Personal correspondence to the author dealing with the genesis of the Mann-Whitney test.

WILLIAM EDWARDS DEMING. 1900–1993. 1994. Alexandria, VA: American Statistical Association. [A pamphlet prepared in support of the W. Edwards Deming Fund]

ZABELL, SANDY. 1994. A conversation with William Kruskal. *Statistical Science* 9:285.

——. 1989. R. A. Fisher and the history of inverse probability. *Statistical Science* 4:247.

Other books and articles, material from which was used in this book:

ANDREWS, D. F., Bickel, P. J., Hampel, F. R., Huber, P. J., Rogers, W. H., and Tukey, J. W. 1972. *Robust Estimates of Location: Survey and Advances*. Princeton, NJ: Princeton University Press.

BARLOW, R. E., Bartholomew, D. J., Bremner, J. M., and Brunk, H. D. 1972. *Statistical Inference Under Order Restrictions: The Theory and Application of Isotonic Regression*. New York: John Wiley & Sons.

BOSECKER, R. R., Vogel, F. A.,Tortora, R. D., and Hanuschak, G. A. 1989. *The History of Survey Methods in Agriculture (1863–1989)*. Washington, DC: U.S. Department of Agriculture, National Agricultural Statistics Service.

BOX, G. E. P., and Tiao, G. C. 1973. *Bayesian Inference and Statistical Analysis*. Reading, MA: Addison-Wesley.

BRESLOW, N. E. 1996. Statistics in epidemiology: The case-control study. *Journal of the American Statistical Association* 91:14.

COCHRAN, WILLIAM G., and Cox, Gertrude M. 1950. *Experimental Designs*. New York: John Wiley & Sons.

COHEN, L. JONATHAN. 1989. *An Introduction to the Philosophy of Induction and Probability*. Oxford: Clarendon Press.

——. 1977. *The Probable and the Provable*. Oxford: Clarendon Press.

CORNFIELD, J., Haenszel, W., Hammond, E. C., Lilienfeld, A. M., Shimkin, M. B., and Wynder, E. L. 1959. Smoking and lung cancer: Recent evidence and a discussion of some questions. *Journal of the National Cancer Institute* 22:173.

DAVID, F. N., and Johnson, N. L. 1951. The effect of non-normality on the power function of the F-test in the analysis of variance. *Biometrika* 38:43.

DAVIES, BRIAN. 1999. *Exploring Chaos—theory and experiment*. Reading, MA: Perseus Books.

DAVIS, PHILIP I. 1980. Are there coincidences in mathematics? *American Mathematical Monthly* 88:311.

DEMING, W. EDWARDS. 1974. Selected topics for the theoretical statistician: Invited talk presented at the Princeton meeting of the Metropolitan Section of the American Society for Quality Control, November 17, 1974.

DOLL, RICHARD, and Hill, Austin Bradford. 1964. Mortality in relation to smoking: Ten years' observations of British doctors. *British Medical Journal* 1:1399.

DOOB, J. L. 1953. *Stochastic Processes*. New York: John Wiley & Sons.

DORN, HAROLD F. 1959. Some problems arising in prospective and retrospective studies of the etiology of disease. *New England Journal of Medicine* 261:571.

EFRON, BRADLEY. 1971. Does an observed sequence of numbers follow a simple rule? (Another look at Bode's law). *Journal of the American Statistical Association* 66:552.

ELDERTON, WILLIAM PALIN, and Johnson, Norman Lloyd. 1969. *Systems of Frequency Curves*. London: Cambridge University Press.
FEINSTEIN, ALVAN R. 1989. Epidemiologic analyses of causation: The unlearned scientific lessons of randomized trials. *Journal of Clinical Epidemiology* 42:481.
FIENBERG, STEPHEN, ed. 1989. *The Evolving Role of Statistical Assessments as Evidence in the Court*. New York: Springer-Verlag.
FISHER, R. A. 1935. *The Design of Experiments*. Subsequent eds. 1937–1966. It was also trans. into Italian, Japanese, and Spanish. Edinburgh: Oliver and Boyd.
——. 1930. *The Genetical Theory of Natural Selection*. Oxford: University Press.
——. 1925. *Statistical Methods for Research Workers*. Subsequent eds. 1928–1970. It was also trans. into French, German, Italian, Japanese, Spanish, and Russian. Edinburgh: Oliver and Boyd.
FITCH, F. B. 1952. *Symbolic Logic: An Introduction*. New York: The Ronald Press Company.
GREENBERG, B. G. 1969. Problems of statistical inference in health with special reference to the cigarette smoking and lung cancer controversy. *Journal of the American Statistical Association* 64:739.
GUMBEL, E. J. 1958. *Statistics of Extremes*. New York: Columbia University Press.
KEYNES, J. M. 1920. A *Treatise on Probability*. New York: Harper and Row (1962).
KOSKO, B. 1993. *Fuzzy Thinking: The New Science of Fuzzy Logic*. New York: Hyperion.
KUHN, T. 1962. *The Structure of Scientific Revolutions*. Chicago: University of Chicago Press.
MENDEL, GREGOR. 1993. *Gregor Mendel's Experiments on Plant Hybrids: A Guided Study*, edited by Alain F. Corcos and Floyd V. Monaghan. New Brunswick, NJ: Rutgers University Press.
PEARSON, KARL. 1935. On Jewish–Gentile relationships. *Biometrika* 220:32.
SALSBURG, D. S. 1992. *The Use of Restricted Significance Tests in Clinical Trials*. New York: Springer-Verlag.
SAVAGE, L. J. 1954. *The Foundations of Statistics*. New York: John Wiley & Sons.

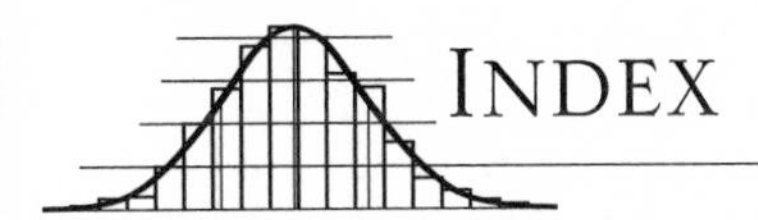

INDEX